T0133117

ELECTRONICS AND COMMUNICATIONS ENGINEERING

Applications and Innovations

ELECTRONICS AND COMMUNICATIONS ENGINEERING

Applications and Innovations

Edited by
T. Kishore Kumar, PhD
Ravi Kumar Jatoth, PhD
V. V. Mani, PhD

Apple Academic Press Inc.
3333 Mistwell Crescent
Oakville, ON L6L 0A2 Canada

Apple Academic Press Inc.
1265 Goldenrod Circle NE
Palm Bay, Florida 32905 USA

Library and Archives Canada Cataloguing in Publication

Title: Electronics and communications engineering : applications and innovations / T. Kishore Kumar, PhD, Ravi Kumar Jatoth, PhD, V.V. Mani, PhD.

Names: National Conference on Electronics and Communication Engineering: Applications and Innovations (2016 : Warangal, India) | Kumar, T. Kishore, editor. | Jatoth, Ravi Kumar, editor. | Mani, V. V., editor.

Description: This book reports the proceedings of the National Conference on Electronics and Communication Engineering: Applications and Innovations, held at the Department of Electronics and Communication Engineering (E.C.E.), National Institute of Technology, Warangal India, on October 21-22, 2016. | Includes bibliographical references and index.

Identifiers: Canadiana (print) 20189069279 | Canadiana (ebook) 20189069287 | ISBN 9781771886932 (hardcover) | ISBN 9781351136822 (PDF)

Subjects: LCSH: Telecommunication systems—Congresses. | LCSH: Wireless communication systems—Congresses. | LCSH: Electrical engineering—Congresses. | LCSH: Microwave antennas—Congresses. | LCSH: Integrated circuits—Very large scale integration—Congresses. | LCSH: Embedded computer systems—Congresses. | LCSH: Signal processing—Digital techniques—Congresses.

Classification: LCC TK5101.A1 N38 2019 | DDC 621.382—dc23

Library of Congress Cataloging-in-Publication Data

Names: Kumar, T. Kishore, editor. | Jatoth, Ravi Kumar, editor. | Mani, V. V., editor.

Title: Electronics and communications engineering : applications and innovations / edited by T. Kishore Kumar, Ravi Kumar Jatoth, V. V. Mani.

Description: Toronto ; New Jersey : Apple Academic Press, 2019. | Includes bibliographical references and index.

Identifiers: LCCN 2018058767 (print) | LCCN 2018061740 (ebook) | ISBN 9781351136822 (ebook) | ISBN 9781771886932 (hardcover : alk. paper)

Subjects: LCSH: Telecommunication systems--Design and construction. | Telecommunication systems--Mathematical models. | Telecommunication systems--Equipment and supplies. | Electrical engineering.

Classification: LCC TK5102.5 (ebook) | LCC TK5102.5 .E42 2019 (print) | DDC 621.382--dc23

LC record available at https://lccn.loc.gov/2018058767

ABOUT THE EDITORS

T. Kishore Kumar, PhD

T. Kishore Kumar, PhD, is currently working as a **Professor** of the Department of Electronics and Communication Education (ECE) at the National Institute of Technology (NIT) in Warangal, India. His research interests include speech signal processing, adaptive signal processing, real time signal processing using embedded systems, and advanced digital system design. He is currently working on two R&D projects sponsored by the Ministry of Human Resource Development (MHRD) and the Science and Engineering Research Board (SERB). Over the course of the last five years, he has been published in 20 international journal publications and 16 national and international conference publications.

Dr. Kumar acquired his BTech degree in electronics and communication engineering from Sri Venkateswara University, Tirupati, India, and his MTech degree in digital systems and computer electronics from J.N.T.U. Hyderabad, India. He received his PhD in Digital Signal Processing from JNTU Hyderabad.

Ravi Kumar Jatoth, PhD

Ravi Kumar Jatoth, PhD, is currently working as an **Associate Professor** of Electronics and Communication Education (ECE) at the National Institute of Technology (NIT) in Warangal, India. He has over 12 years of teaching and research experience in control system design, signal processing, and bio-inspired evolutionary algorithms. He has published 20 international journal papers and 30 international conference papers. He is an Associate Editor of the *International Journal of VLSI Design Tools & Technology* (JoVDTT) and Associate Editor of MECS *International Journal of Image, Graphics and Signal Processing* (IJIGSP), Hong Kong.

Professor Kumar Jatoth received his BE degree in Electronics and Communications Engineering from Osmania University in Hyderabad, and his MTech degree in Instrumentation and Control Systems from Jawaharlal Nehru Technological University in Hyderabad. He completed his PhD from the National Institute of Technology in Warangal, India, where he is a current educator.

V. V. Mani, PhD

V. V. Mani, PhD, is currently working as an **Associate Professor** at the National Institute of Technology in Warangal, India. She has over 15 years of teaching and research experience in her specific areas of interest, which include signal processing for communication and smart antenna design. She has published 20 international journal papers and 20 international conference papers.

Professor Mani received her BE and ME degrees in electronics and communications engineering from Andhra University in Vishakhapatnam and earned her PhD from the Indian Institute of Technology Delhi, India.

CONTENTS

CONTRIBUTORS

Akhil
Professor, IV ECE, Department of ECE, Hyderabad Institute of Technology and Management, Hyderabad, Telangana, India

L. Anjaneyulu
Research Scholar, Associate Professor, Department of ECE, National Institute of Technology, Warangal, Telangana, India. E-mail: anjan@nitw.ac.in

Shravan Kumar Bandari
Department of Electronics and Communication Engineering, National Institute of Technology, Warangal, Telangana, India. E-mail: shravnbandari@gmail.com

Prabhu G. Benakop
Principal, Department of ECE, Indur Institute of Engineering and Technology, Ponnala (V), Siddipet, Medak, Telangana, India. E-mail: pgbenakop@ieee.org; pgbenakop@rediffmail.com

R. Bhashya Sri Bharati
Computer Science and Engineering, Nalla Malla Reddy Engineering College, Hyderabad, India. E-mail: bhashyasri@gmail.com

Boda Bhasker
Department of Electrical and Computer Engineering, Wollega University, Nekemte, Ethiopia

D. Durga Bhavani
Department of Computer Science and Engineering, CVR College of Engineering (Autonomous), Hyderabad, Telangana, India
Computer Science and Engineering, Nalla Malla Reddy Engineering College, Hyderabad, India. E-mail: drddurgabhavani@gmail.com

Adupa Chakradhar
Jayamukhi Institute of Technological Sciences, Department of ECE, Warangal 506002, Telangana, India. E-mail:adupa.chakradhar@gmail.com

B. Sai Chakradhar
Department of ECE, VITS, Hyderabad, Telangana, India

K. Shahu Chatrapati
Associate Professor and HOD, Department of CSE, JNTUH College of Engineering Manthani, Karimnagar, Telangana, India. E-mail: shahujntu@gmail.com

V. Devika
Research Scholar, Department of ECE, KL University, Vijayawada, Andhra Pradesh, India; Associate Professor, Department of ECE, Hyderabad Institute of Technology and Management, Hyderabad, Telangana, India

Prateek Dhanuka
Department of Electronics and Communication Engineering, National Institute of Technology, Andhra Pradesh, India

P. V. Hunagund
Department of Applied Electronics, Gulbarga University, Gulbarga 585106, Karnataka, India

Ravi Kumar Jatoth
Department of Electronics and Communication Engineering, National Institute of Technology,
Warangal, Telangana, India

Mounika Kamatam
Computer Science and Engineering, Nalla Malla Reddy Engineering College, Hyderabad, India.
E-mail: mounikakamatam25@gmail.com

Eduru Hemanth Kumar
Department of Electronics and Communication Engineering, National Institute of Technology,
Warangal, Telangana, India. E-mail: ehemanth21@gmail.com

I. Hemanth Kumar
Department of ECE, NIT, Warangal, Telangana, India

T. Lalith Kumar
Annamacharya Institute of Technology & Sciences, C.K. Dinne, Kadapa 516003, Andhra Pradesh,
India

K. Sarat Kumar
Associate Professor, Department of ECE, Hyderabad Institute of Technology and Management,
Hyderabad, Telangana, India

Gaddam Shravan Kumar
Jayamukhi Institute of Technological Sciences, Department of ECE, Warangal 506002, Telangana,
India. E-mail: shravankumar2014@gmail.com

K. Pruthvi Krishna
Department of Electronics and Communication Engineering, National Institute of Technology,
Warangal, Telangana, India. E-mail: pruthvikurada@gmail.com

M. N. S. Lahari
UG Student, Department of ECE, Sri Chandrasekharendra Saraswathi Vishwa Mahavidyalaya,
Kancheepuram, Tamil Nadu, India. E-mail: mns.lahari@gmail.com

P. Venu Madhav
Research Scholar, Department of ECE, KL University, Guntur, Andhra Pradesh, India.
E-mail: venu7485@gmail.com

V. V. Mani
Department of Electronics and Communication Engineering, National Institute of Technology,
Warangal, Telangana, India. E-mail: vvmani@nitw.ac.in

Sagar Mekala
Research Scholar, JNTUH, Hyderabad, India
Assistant Professor (C), Department of CSE, UCET, Mahatma Gandhi University, Nalgonda,
Telangana, India. E-mail: arjunnannahi5@hotmail.com

Nihar Ranjan Panda
Department of ECE, SITAM, Vizianagaram, Andhra Pradesh, India

Krishna Patteti
Associate Professor, Department of Electronics and Communication Engineering, Jayamukhi Institute
of Technological Sciences, Warangal, Telangana, India. E-mail: kpatteti@gmail.com

D. Prabhakar
Department of ECE, Gudlavalleru Engineering College (A), Seshadri Rao Knowledge Village,
Gudlavalleru, Andhra Pradesh, India

Pragnya
IV ECE, Department of ECE, Hyderabad Institute of Technology and Management, Hyderabad,
Telangana, India

K. Nunny Praisy
M.Tech-VLSID, SVECW (Autonomous), Bhimavaram, Andhra Pradesh, India.
E-mail: knpraisy@gmail.com

M. Sivaganga Prasad
Department of Electronics and Communication, KKR and KSR Institute of Technology and Sciences,
Guntur, Andhra Pradesh, India. E-mail: msivagangaprasad@gmail.com

C. S. Preetham
Department of Electronics and Communication Engineering, KL University, Guntur,
Andhra Pradesh, India. E-mail: cspreetham@kluniversity.in

A. Selwin Mich Priyadharson
Vel Tech Rangarajan Dr Sagunthala R&D Institute of Science and Technology, Avadi, Chennai,
Tamil Nadu, India. E-mail: aselwinmich@veltechuniv.edu.in

Kotipalli Pushpa
Professor, Department of ECE, Shri Vishnu Engineering College for Women, Bhimavaram,
Andhra Pradesh, India. E-mail: pushpak@svecw.edu.in

Vaibhav Singh Rajput
Department of Materials and Metallurgical Engineering, National Institute of Technology,
Andhra Pradesh, India

K. S. N. Raju
Assistant Professor, SVECW (Autonomous), Bhimavaram, Andhra Pradesh, India

T. V. Ramakrishna
Department of Electronics and Communication Engineering, KL University, Guntur, Andhra Pradesh,
India

B. Ramesh
Department of Electronics and Communication Engineering, Kamala Institute of Technology &
Science, Huzurabad 505468, Telangana, India. E-mail: brameshb2@rediffmail.com

M. Asha Rani
Department of Electronics and Communication Engineering, JNTUH College of Engineering,
Hyderabad 500085, Telangana, India. E-mail: ashajntu1@yahoo.com

Swetha Ravikanti
Department of ECE, Vardhaman College of Engineering, Hyderabad, Telangana, India
Research Scholar, Department of ECE, National Institute of Technology, Warangal, Telangana, India.
E-mail: swetha.rks@nitw.ac.in

M. Sampath Reddy
Associate Professor, Department of Electronics and Communication Engineering,
Jayamukhi Institute of Technological Sciences, Warangal, Telangana, India.
E-mail: sampath_reddy@srecwarangal.ac.in

K. Jaya Shankar Reddy
Department of Computer Science and Engineering, National Institute of Technology,
Andhra Pradesh, India

Kanaparthi Revathi
PG Scholar, Department of ECE, Shri Vishnu Engineering College for Women, Bhimavaram, Andhra Pradesh, India. E-mail: revathikanaparthi@gmail.com

M. V. Ganeswara Rao
Associate Professor, Department of ECE, SVECW, Bhimavaram, Andhra Pradesh, India

P. Mallikarjuna Rao
Department of ECE, AUCE (A), Andhra University, Visakhapatnam, Andhra Pradesh, India. E-mail: pmraoauece@yahoo.com

K. Srinivasa Rao
Professor, Department of Electronics and Communication Engineering, TRR Engineering College, Hyderabad, Telangana, India

V. Saidulu
Associate Professor, Department of ECE, MGIT, Hyderabad 500075, Telangana, India. E-mail: saiduluvadtya@gmail.com

P. N. S. Sailaja
Department of ECE, SITAM, Vizianagaram, Andhra Pradesh, India

P. Sandeep
Department of ECE, VITS, Hyderabad, Telangana, India. E-mail: san9ap@gmail.com

B. Saroja
Vel Tech Rangarajan Dr Sagunthala R&D Institute of Science and Technology, Avadi, Chennai, Tamil Nadu, India. E-mail: boda.saroja@gmail.com

Rupali Satapathy
Department of EEE, HIT, Bhubaneswar, Odisha, India

Bharatha Sateesh
Assistant Professor, Department of ECE, Vaagdevi College of Engineering, JNTU, Warangal, Telangana, India. E-mail: basateesh27@gmail.com

K. Satisha
PG Scholar, Department of ECE, SVECW, Bhimavaram, Andhra Pradesh, India. E-mail: kodi.satisha@gmail.com

M. Satyanarayana
Department of ECE, MVGR College of Engineering (A), Vizianagaram, Andhra Pradesh, India. E-mail: profmsn26@gmail.com

Md. Sharuque
Department of ECE, VITS, Hyderabad, Telangana, India

N. Shehanaz
Annamacharya Institute of Technology & Sciences, C.K. Dinne, Kadapa 516003, Andhra Pradesh, India. E-mail: shehanazbegum467@gmail.com

F. B. Shiddanagouda
Sphoorthy Engineering College, Hyderabad 501510, Telangana, India. E-mail: siddu.kgp09l@gmail.com

K. V. Sridhar
Department of ECE, NIT, Warangal, Telangana, India

K. Ch. SriKavya
Professor, Department of ECE, KL University, Vijayawada, Andhra Pradesh, India

G. R. L. V. N. Srinivasaraju
Professor, Head of the Department, Department of ECE, SVECW, Bhimavaram, Andhra Pradesh, India

Ch. Sulakshana
Department of ECE, Vardhaman College of Engineering, Hyderabad, Telangana, India

Siva Kumara Swamy
Sphoorthy Engineering College, Hyderabad 501510, Telangana, India

Anil Kumar Tipparti
Professor, Department of Electronics and Communication Engineering, CMR Institute of Technology, Hyderabad, Telangana, India

K. Umapathy
Department of ECE, Sri Chandrasekharendra Saraswathi Vishwa Mahavidyalaya, Kancheepuram, Tamil Nadu, India. E-mail: umapathykannan@gmail.com

R. M. Vani
University Science Instrument Center, Gulbarga University, Gulbarga 585106, Karnataka, India

G. Deeshma Venkatakanakadurga
PG Scholar, Department of ECE, SVECW, Bhimavaram, Andhra Pradesh, India.
E-mail: deeshma555@gmail.com

K. Karthikeya Yadav
Department of Computer Science and Engineering, National Institute of Technology, Andhra Pradesh, India

ABBREVIATIONS

ACO	ant colony optimization
ADCs	analog-to-digital conversions
ALNDLDC	ALOHA-like neighbor discovery in low duty cycle
ARM	Advanced RISC Machines
ATS	antenna tracking system
AWGN	additive white Gaussian noise
BC-SRR	broadside-coupled square, ring resonator
BDE	boundary displacement error
BIST	built-in self-test
BS	base station
CCA	clear channel assessment
CCD	colon cancer diagnosis
CD	cooperative diversity
CG	coefficient generator
CGG	Center for Good Governance
CISC	complex instruction set computer
CLA	carry-look-ahead adder
CMOS	complementary metal-oxide-semiconductor
CNC	computer numerical control
CP	circular polarization
CP	cyclic prefix
CPMAs	circularly polarized microstrip antennas
CPW	coplanar waveguide
CPW	coplanar waveguide
CR	cognitive radio
CRC	colorectal cancer
CS	coefficient selector
CSA	carry save addition
CSF	cerebrospinal fluid
CSI	channel state information
CSLA	carry select adder
CSMA/CA	carrier-sense multiple access with collision avoidance
CSRR	complementary split-ring resonators
CUT	circuit under test

DAC	digital-to-analog converter
DE	differential evolution
DFT	discrete Fourier transform
DG	data generator
DIF	decimation in frequency
DSP	digital signal processing
ECCs	envelope correlation coefficients
ED	error distance
ENDP	efficient neighbor discovery protocol
FA	full adder
FAM	fused add–multiply
FCM	fuzzy C means
FCS	frame check sequence
FFT/IFFT	fast Fourier/inverse Fourier transform
FPGAs	field programmable gate arrays
FSR	full-scale range
GCE	global consistency error
GECC	gene expression-based ensemble classification of colon samples
GFDM	generalized frequency division multiplexing
GM	gray matter
GNU	Gnu's Not Unix
GSM	global system for mobile communication
HA	half adder
HFSS	high-frequency structure simulation
IC	integrated circuit
IG	information gain
IOB	input/output block
ISHRAE	Indian Society of Heating Refrigeration & Air Conditioning Engineers
ITAE	integral time absolute error
LC circuit	inductance–capacitance circuit
LFSR	linear feedback shift register
LNA	low-noise amplifier
LSB	least significant bits
LSI	large-scale integration
LTE	long-term evolution
MAC	medium access control
MB	modified booth
MDC	multipath delay commutator

MF	matched filtering
MIMO	multiple input multiple output
MLBC	multilevel boost converter
MPA	microstrip patch antenna
MPP	maximum power point
MPPT	maximum power point tracking
MR	magnetic resonance
MRI	magnetic resonance imaging
MSB	most significant bit
NCAER	National Council for Applied Economics Research
OCSRRs	octagon split-ring resonators
OFDM	orthogonal frequency division multiplexing
ORA	output response analyzer
OTM	on-the-move
P&O	perturb and observe
PCB	printed-circuit-board
PHY	physical
PIC	peripheral interface controller
PID	proportional-integral derivative
PIFA	planar inverted-F antenna
PN	pseudo noise
PNDLPL	passive neighborhood discovery for low power listening MAC protocols
PRPG	pseudorandom pattern generator
PRR	packet-received ratio
PSNR	peak signal-to-noise ratio
PSO	particle swarm optimization
PUs	primary users
PV	photovoltaic
QCA	quantum dot cellular automata
RCA	ripple carry adder
RF	radio frequency
RFID	radio frequency identification
RI	Rand index
RISC	reduced instruction set computer
RRC	root raised cosine
RTL	register transfer logic
RTL	register-transfer level
SATCOM	satellite communications
SCH	schematic

SDC	single-path delay commutator
SDF	single-path delay feedback
sDNA	stool DNA
SDR	software-defined radio
SIMWS	substrate-integrated metallic wall structure
SMB	sum-to-modified booth
SNR	signal-to-noise ratio
SOTM	satellite communications on the move
SRR	split-ring resonators
SU	secondary users
TEM waveguide	transverse electromagnetic waveguide
TLBO	teacher learning-based optimization
TPG	test pattern generator
TVIW	time-varying initial weight
UHD	USRP hardware driver
USPR	universal software radio peripheral
VHDL	very high speed integrated hardware description language
VLSI	very-large-scale integration
VNA	vector network analyze
VoCA	voice output communication aid
VoI	variation of information
VSWR	voltage standing wave ratio
WiMAX	worldwide interoperability for microwave access
WLAN	wireless local area network
WM	white matter
WSN	wireless sensor network
XOR	exclusive-OR

PREFACE

This book reports the proceedings of the National Conference on Electronics and Communication Engineering: Applications and Innovations, held at the Department of Electronics and Communication Engineering (E.C.E.), National Institute of Technology, Warangal India, on October 21–22, 2016. The purpose of this conference was to explore the research advances in the field of wireless communication, signal processing, embedded systems, VLSI, microwave, and antennas. This book provides insights of present technological challenges that help to induce new ideas for future research and collaboration to graduate students, researchers, and professionals.

This book is organized as follows:

Part I: Microwave and Antennas
Part II: Communication Systems
Part III: Very Large-Scale Integration
Part IV: Embedded Systems
Part V: Intelligent Control and Signal Processing Systems

Only those papers that have undergone a rigorous review process, followed by a plagiarism check using Turnitin software, were accepted, and the revised manuscripts were published in the conference proceedings.

We thank all the participants for their contribution to the conference, which has helped to prepare this book. We also express our sincere thanks to a number of research scholars who helped in preparing the content of this book. The assistance of all the individuals and contributors is greatly appreciated by the authors.

PART I
Microwave and Antennas

CHAPTER 1

SIERPINSKI DIAMOND FRACTAL ANTENNA ARRAY USING A MITERED BEND FEED NETWORK FOR MULTIBAND APPLICATIONS

D. PRABHAKAR[1*], P. MALLIKARJUNA RAO[2], and M. SATYANARAYANA[3]

1Department of ECE, Gudlavalleru Engineering College (A), Seshadri Rao Knowledge Village, Gudlavalleru, Andhra Pradesh, India

2Department of ECE, AUCE (A), Andhra University, Visakhapatnam, Andhra Pradesh, India

3Department of ECE, MVGR College of Engineering (A), Vizianagaram, Andhra Pradesh, India

Corresponding author. E-mail: prabhakar.dudla@gmail.com

ABSTRACT

Design and performance measures of a Sierpinski diamond fractal Antenna array with mitered bend feed network for multiband applications are proposed in this paper. To achieve wideband/multiband antennas, one technique is by applying fractal shapes into antenna geometry. These antennas are designed using HFSS on FR4 substrate with dielectric constant of 4.4 and fed with 50 Ω microstrip line. Diamond antenna array has been fabricated and tested using a vector network analyzer (VNA) and performance of the proposed antenna is confirmed by using simulation and experimental results.

1.1 INTRODUCTION

Fractal geometries in antenna design have been of particular interest in recent years due to its suitability for compact personal communication equipment.

These geometries have two common properties: *space-filling* and *self-similarity*. While the space-filling property is used to reduce the antenna size,[1-4] the self-similarity property can be successfully applied to design multiband fractal antennas. In particular, an antenna with self-similar structures provides similar surface current distributions for different frequencies, which leads to multiband behavior.[5-7] A fractal antenna can be explained as an antenna that uses a fractal design to maximize the length of material that transmits or receives electromagnetic signals within a given total surface area. Due to this reason, fractal antennas are very compact and hence are anticipated to have useful applications in cellular telephone and microwave communications. Fractal antenna's response differs markedly from traditional antenna designs, in the sense that it is capable of operating optimally at many different frequency ranges simultaneously. Fractal antennas are antennas that have the shape of fractal structures. The fractal antennas consist of geometrical shapes that are repeated. Each one of the shapes has unique attributes. There are many fractal geometries such as Sierpinski gasket, Sierpinski carpet, Koch Island, Hilbert curve, and Miskowski.[8-11]

1.1.1 SIERPINSKI DIAMOND

Sierpinski diamond fractal antenna is the widely studied fractal geometry[3] for antenna application. The fractal antenna consists of geometrical shapes that are repeated. Each one of these has unique attributes. The self-similarity that is distributed on this antenna is expected to cause its multiband characteristics. On the other hand, it can solve a traditional antenna that operates at single frequency. In this chapter, first 3-iterated diamond fractal patch antenna has bestowed and supported the Sierpinski gasket fractal geometry as shown in Figure 1.1.

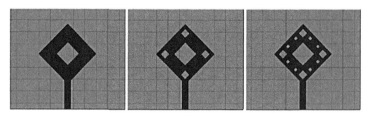

Stage 1   stages 2   stage 3

FIGURE 1.1   Sierpinski diamond antenna with third iteration.

1.2   DESIGN OF FEED NETWORK FOR AN ARRAY

In the farthest communications, antennas with high directivity are regularly required. Single-element antenna is not suitable for high gain or high directivity. High gain can be achieved by an assemblage of antennas, called an array. In the construction of an array, feed network design is essential. Feed network is used in an array to regulate the amplitude and phase of the radiating elements to control the beam scanning properties. Thus, in selecting and optimizing the feed network, the design of an array is crucial. Different types of feed networks are parallel feed, T-split power divider, quarter-wave transformer, and mitered bend feed.

1.2.1   MITERED BEND FEED NETWORK

The transmission lines in the feed networks have many bends to guide the signals to/from the elements. A 90° bend in a microstrip line produces a large reflection from the end of the line. Some signal bounces around the corner, but a large portion reflects back the way the signal traveled down the line. If the bend is an arc of radius at least three times the strip width, then reflections are minimal. This large bend takes up a lot of real estate compared to the 90° bend. A sharp 90° bend behaves as a shunt capacitance between the ground plane and the bend. To create a better match, the bend is mitered to reduce the area of metallization and remove the excess capacitance. The signal is no longer normally incident to microstrip edge, so it reflects from the end down the other arm. Figure 1.2 shows a straight bend (90° bend) and Figure 1.3 shows a mitered bend.

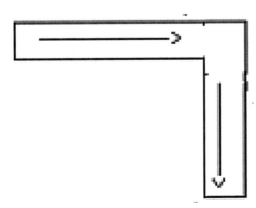

FIGURE 1.2   90° bend.

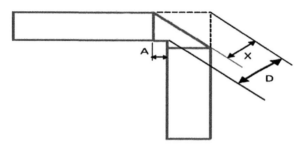

FIGURE 1.3 45° miter bend.

Design requires T-junction and miter bending (MBEND) modification to generate low insertion loss at the input port. Mitered T-junction and microstrip bends were applied to have low reflection and insertion losses. The mitered T-junction of 3 dB power divider is shown in Figure 1.4.

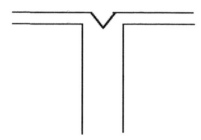

FIGURE 1.4 Mitered "T" bend.

1.3 ANTENNA DESIGN

A schematic diagram of the diamond microstrip patch antenna in a very Sierpinski carpet kind is shown in Figure 1.5. A scaling factor of $\delta = 1/3$ was chosen to maintain the perfect geometry symmetry of fractal structure. A diamond-shaped structure of dimensions 50 mm × 50 mm has been designed. With the scaling factor of 2.5, again a diamond-shaped structure with dimensions of 20 mm × 20 mm has been designed and iterated on the existing main diamond structure. This is the first iteration. During the second iteration, four diamond shapes of dimensions 8 mm × 8 mm have been designed and iterated on the main diamond structure. In the third iteration, eight more diamond shapes of dimensions 4 mm × 4 mm have been designed and iterated. The substrate used to design the proposed fractal antenna is FR-4 of thickness $d = 1.6$ mm with relative permittivity $\varepsilon_r = 4.4$.

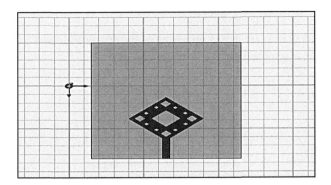

FIGURE 1.5   Structure of single-element Sierpinski diamond fractal antenna.

The dimensions of the Sierpinski fractal antenna can be approximately calculated as follows[12–16]:

A. Calculation of width (W):

Width of the patch antenna is calculated by using

$$W = \frac{c}{2f_0\sqrt{(\varepsilon_r + 1/2)}} \tag{1.1}$$

where $c = 3 \times 10^8$ m/s

B. Calculation of actual length (L):

The effective length of patch antenna depends on the resonant frequency (f_0).

$$L_{\text{eff}} = \frac{c}{2f_0\sqrt{\varepsilon_{\text{reff}}}} \tag{1.2a}$$

where

$$\varepsilon_{\text{reff}} = \frac{\varepsilon_r + 1}{2} + \frac{\varepsilon_r - 1}{2}\left[1 + 12\frac{h}{W}\right]^{-1/2} \tag{1.2b}$$

Actual length and effective length of a patch antenna can be related as

$$L = L_{\text{eff}} - 2\Delta L \tag{1.3}$$

where ΔL is a function of effective dielectric constant $\varepsilon_{\text{reff}}$ and the width-to-height ratio (W/h)

$$\frac{\Delta L}{h} = 0.412\frac{(\varepsilon_{\text{reff}} + 0.3)\big((W/h) + 0.264\big)}{(\varepsilon_{\text{reff}} - 0.258)\big((W/h) + 0.8\big)} \tag{1.4}$$

C. *Calculation of feed width (W_f):*

To achieve 50 Ω characteristic impedance, the required feed width-to-height ratio (W_f/h) is computed as

$$\frac{W_f}{h} = \begin{cases} \dfrac{8e^A}{e^{2A}-2} \dfrac{W_0}{h} \leq 2 \\ \dfrac{2}{\pi}\left\{B-1-\ln(2B-1)+\dfrac{\varepsilon_r-1}{2\varepsilon_r}\left[\ln(B-1)+0.39-\dfrac{0.61}{\varepsilon_r}\right]\right\}\dfrac{W_0}{h} \geq 2 \end{cases}$$ (1.5a)

where

$$A = \frac{Z_0}{60}\sqrt{\frac{\varepsilon_r+1}{2}} + \frac{\varepsilon_r+1}{\varepsilon_r-1}\left(0.23+\frac{0.11}{\varepsilon_r}\right)$$ (1.5b)

$$B = \frac{377\pi}{2Z_0\sqrt{\varepsilon_r}}$$ (1.5c)

D. *Miter bend designed equation (D):* $D = \mathbf{D = w\sqrt{2}}$

$$X = W\sqrt{2}\times\left(0.52+0.65\times e^{\left(-1.35\times(W/h)\right)}\right)$$ (1.6)

$$A = X\sqrt{2}-W$$

E. *The number of iterations is*

$$N_n = 8^n$$ (1.7)

F. *The ratio of fractal length is*

$$L_n = \left(\frac{1}{3}\right)^n$$ (1.8)

G. *The ratio for the fractal area after the nth iteration is*

$$A_n = \left(\frac{8}{9}\right)^n$$ (1.9)

where *n* is iteration *n*th stage number.

H. *The antenna is matched approximately at frequencies*

$$f_n \approx 0.26\frac{c}{h}\delta^n$$ (1.10)

where $\delta = 3$ is the log-period constant, n is a natural number, c is the speed of light in vacuum, and h is the height of the largest gasket.

1.4 RESULTS

1.4.1 DIMENSIONS OF SIERPINSKI DIAMOND FRACTAL ANTENNA

Two-element Sierpinski diamond antenna array with mitered bend feed network is designed and fabricated with the design equations and is shown in Figures 1.6 and 1.7. Dimensions of single-element Sierpinski diamond fractal antenna are presented in Table1.1. Simulated and practical results such as reflection coefficient, voltage standing wave ratio (VSWR), and gain are observed in Figures 1.8–1.12, respectively. The obtained results are mentioned in Table 1.2.

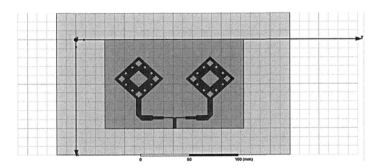

FIGURE 1.6 Geometry for two-element antenna array of mitered bend feed network.

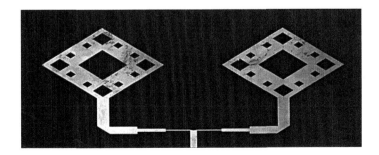

FIGURE 1.7 Fabricated patch of two-element antenna array of mitered bend feed network.

TABLE 1.1   Dimensions of Single-element Sierpinski Diamond Fractal Antenna.

Parameter	Length (mm)
L1	35.355
L2	14.14
L3	5.656
L4	2.828
W1	5
A	24

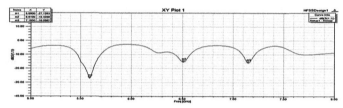

FIGURE 1.8   Reflection coefficient curve of the two-element antenna array of mitered bend feed network.

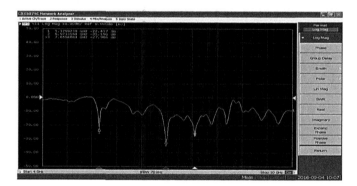

FIGURE 1.9   Reflection coefficient curve of fabricated two-element antenna array of mitered bend feed network.

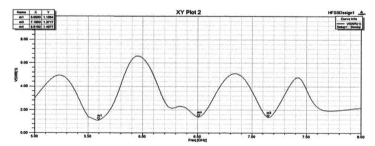

FIGURE 1.10   VSWR curve of the two-element antenna array of mitered bend feed network.

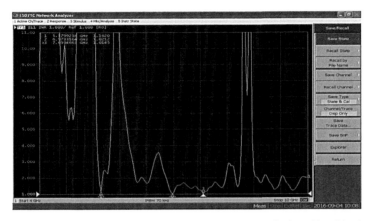

FIGURE 1.11   VSWR of fabricated two-element antenna array of mitered bend feed network.

FIGURE 1.12   (See color insert.) Gain plot of the two-element antenna array of mitered bend feed network.

TABLE 1.2   Results of Two-element Sierpinski Diamond Antenna Array with Mitered Bend Feed Network.

Resonant frequency (GHz)	Simulated	5.59
		6.5
		7.15
	Measured	5.37
		6.97
		7.65
Reflection coefficient (S11) (dB)	Simulated	−27.12 at 5.59 GHz
		−15.42 at 6.5 GHz
		−16.09 at 7.15 GHz
	Measured	−22.43 at 5.37 GHz
		−31.19 at 6.97 GHz
		−27.96 at 7.65 GHz

TABLE 1.2 *(Continued)*

VSWR	Simulated	1.10 at 5.6 GHz
		1.4 at 6.5 GHz
		1.3 at 7.3 GHz
	Measured	1.6 at 5.17 GHz
		1.02 at 6.9 GHz
		1.01 at 7.6 GHz
Gain (dB)	Simulated	6.00

1.5   CONCLUSION

In this chapter, a microstrip feed Sierpinski diamond fractal antenna array is designed and implemented by using mitered bend feed network. Then, an improvement is observed in gain from 4.952 dB (single element) to 6.00 dB (two elements). The linear/circular polarization behavior has also been achieved from the proposed structure. The proposed antenna can be used for various wireless communication applications. Also, it has been found that as iteration increases in fractal structure, number of bands also increases. The simulated and measured results are found to be in good agreement. It can be summed up that multibands are developed besides the resonance frequency. For further improvement in gain, we have to increase the number of elements in the array.

KEYWORDS

- **microstrip antenna**
- **Sierpinski diamond fractal antenna**
- **mitered bend feed network**
- **reflection coefficient**
- **HFSS**

REFERENCES

1. Gianvittorio, J. P.; Rahmat-Samii, Y. Fractal Antenna: A Novel Antenna Miniaturization Technique and Applications. *IEEE Antennas Propag. Mag.* **2002,** *44* (1), 20–36.
2. Best, S. R. On the Significance of Self-similar Fractal Geometry in Determining the Multiband Behavior of the Sierpinski Gasket Antenna. *IEEE Antennas Wireless Propag. Lett.* **2002,** *1* (1), 22–25.
3. Mandelbrot, B. B. *The Fractal Geometry of Nature*; W.H. Freeman and Company: New York, 1983.
4. Puente, C.; Romeu, J.; Pous, R.; Ramis, J.; Hijazo, A. Small but Long Koch Fractal Monopole. *IEEE Electron. Lett.* **1998,** *34* (1), 9–10.
5. Borja, C.; Romeu, J. Multiband Sierpinski Fractal Patch Antenna. *Antennas Propag. Soc. Int. Symp., IEEE* **2000,** *3*, 1708–1711.
6. Harris, J. W.; Stocker, H. Scaling Invariance and Self-similarity and Construction of Self-similar Objects. *Handbook of Mathematics and Computational Science*; Springer-Verlag: New York, 1998; Sec. 4.11.1–4.11.2, p. 113.
7. Puente-Baliarda, C.; Romeu, J.; Pous, R.; Cardama, A. On the Behavior of the Sierpinski Multiband Fractal Antenna. *IEEE Trans.* **1998,** *46* (4), 517–524.
8. Balanis, C. A. *Antenna Theory: Analysis and Design*, 3rd ed.; John Wiley & Sons, Inc.: Hoboken, NJ, 2005.
9. Rumsey, V. H. *Frequency Independent Antenna*; Academic Press: New York, London, 1966.
10. Cohen, N. L. Fractal Antennas Part 1. *Commun. Q.* **1995,** *Summer*, 5–23.
11. Hohlfeld, R. G.; Cohen, N. L. Self-similarity and the Geometric Requirements for Frequency Independence in Antennas. *Fractals* **1999,** *7*, 79–84.
12. Khan, S. N.; Hu, J.; Xiong, J.; He, S. Circular Fractal Monopole Antenna for Low VSWR UWB Applications. *Progress Electromagn. Res. Lett.* **2008,** *1*, 19–25.
13. Werner, D. H.; Ganguly, S. An Overview of Fractal Antenna Engineering Research. *IEEE Antennas Propag. Mag.* **2003,** *45* (1), 38–57.
14. Garg, R. *Micro Strip Antenna Design Handbook*; Artech House: Boston, MA, 2001.
15. Prabhakar, D.; Mallikarjuna Rao, P.; Satyanarayana, M. Design and Performance Analysis of Micro Strip Antenna Using Different Ground Plane Techniques for WLAN Application. *Int. J. Wireless Microwave Technol.* **2016,** *6* (4), 48–58.
16. Prabhakar, D.; Mallikarjuna Rao, P.; Satyanarayana, M. Characteristics of Patch Antenna with Notch Gap Variation for Wi-Fi Application. *Int. J. Appl. Eng. Res.* **2016,** *11* (8), 5741–5746.

CHAPTER 2

SIERPINSKI DIAMOND FRACTAL ANTENNA ARRAY USING A QUARTER-WAVE FEED NETWORK FOR WIRELESS APPLICATIONS

D. PRABHAKAR[1*], P. MALLIKARJUNA RAO[2], and M. SATYANARAYANA[3]

[1]*Department of ECE, Gudlavalleru Engineering College (A), Seshadri Rao Knowledge Village, Gudlavalleru, Andhra Pradesh, India*

[2]*Department of ECE, AUCE (A), Andhra University, Visakhapatnam, Andhra Pradesh, India*

[3]*Department of ECE, MVGR College of Engineering (A), Vizianagaram, Andhra Pradesh, India*

Corresponding author. E-mail: prabhakar.dudla@gmail.com

ABSTRACT

To achieve wideband/multiband antennas, one technique is by applying fractal shapes into antenna geometry. The design of Sierpinski diamond fractal antenna array with quarter wave feeding technique has been presented in this paper. The antenna is designed and simulated by using HFSS software. FR4 glass epoxy having thickness 1.6 mm with dielectric constant 4.4 is used as a substrate material for the designing of proposed antenna. The different parameters of designed antenna are calculated such as reflection coefficient, voltage standing wave ratio, gain, and radiation pattern. The designed antenna is fabricated and measured using VNA (vector network analyzer). On comparison, it shows that the measured results of fabricated antennas are in good agreement with simulated results.

2.1 INTRODUCTION

In the present scenario, microstrip patch antenna is gaining much attention in the multifunctional wireless communication system such as WLAN
(wireless local area network), satellite, mobile, radar, and biomedical
systems.[1] These systems require an antenna with high gain, large impedance bandwidth, and good radiation pattern throughout the entire operating
frequency bands.[2] Different techniques have been developed to obtain
multiband antennas with compact size. We can find fractal antennas
with different geometries (Sierpinski gasket, Sierpinski carpet, and Koch
curves)[3–5] and planar inverted-F antenna (PIFA).[6,7] Fractal geometries in
antenna design have been of particular interest in recent years due to its
suitability for compact personal communication equipment. These geometries have two common properties: *space-filling* and *self-similarity*. While
the space-filling property is used to reduce the antenna size,[8–11] the self-
similarity property can be successfully applied to design multiband fractal
antennas. In particular, an antenna with self-similar structures provides
similar surface current distributions for different frequencies, which leads
to multiband behavior.[12–14]

2.1.1 SIERPINSKI DIAMOND

Sierpinski diamond fractal antenna is the widely studied fractal geometry
for antenna application. The fractal antenna consists of geometrical shapes
that are repeated. Each one of these has unique attributes. The self-similarity
that is distributed on this antenna is expected to cause its multiband characteristics. On the other hand, it can solve a traditional antenna that operates
at single frequency. In this chapter, first 3-iterated diamond fractal patch
antenna has bestowed and supported the Sierpinski gasket fractal geometry
as shown in Figure 2.1.

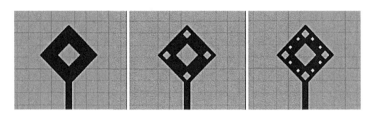

FIGURE 2.1   Sierpinski diamond antenna with third iteration.

2.2   DESIGN OF FEED NETWORK FOR AN ARRAY

In the farthest communications, antennas with high directivity are regularly required. Single element antenna is not suitable for high gain or high directivity. High gain can be achieved by an assemblage of antennas, called an array. In the construction of an array, feed network design is essential. Feed network is used in an array to regulate the amplitude and phase of the radiating elements to control the beam scanning properties. Thus, in selecting and optimizing the feed network, the design of an array is crucial. Different types of feed networks are series feed, parallel feed, T-split power divider, quarter-wave transformer, and mitered bend feed.

2.2.1   QUARTER-WAVE TRANSFORMER FEED

The transmission lines such as coaxial cables, strip lines, and microstrip lines are used in making most feed networks. Impedance matching is vital for ensuring efficient power transfer through feed network. The use of quarter-wave transformer is the best solution for achieving impedance matching. The reflection coefficient between two impedances Z1 and Z3 is cut down by a matching circuit. The quarter-wave transformer shown in Figure 2.2 is one of the most commonly used matching circuits. At the center frequency, a section of transmission line $\lambda/4$ (the quarter-wave transformer) long placed between the two transmission lines eliminates the reflection coefficient if its impedance is $Z_2 = \sqrt{Z_1 Z_3}$

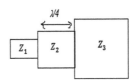

FIGURE 2.2   Quarter-wave transformer.

A two-element corporate-fed microstrip uniform array is illustrated in Figure 2.3. The tree-like structure of the feed appropriately combines/distributes the signals from/to the elements. With a view to matching the lines of different impedances, a quarter-wave transformer appears at the splits. The 50-Ω input line splits into two 100 Ω lines. If the microstrip line continues to split like this, then the lines feeding the elements would be 200 Ω, 400 Ω, and so on. Then, the microstrip line will be very thin. Besides this, the element impedance should be very high for matching. But the element is fed with inset feed; however, the transmission line will have an impedance of 50 Ω. Thus, the 100-Ω line is converted back to 50 using a quarter-wave transformer of 70.7 Ω.

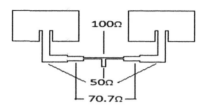

FIGURE 2.3   Two-element corporate-fed microstrip array with quarter-wave transformer.

2.3   ANTENNA DESIGN

A schematic diagram of the diamond microstrip patch antenna in a very Sierpinski carpet kind is shown in Figure 2.4. A scaling factor of $\delta = 1/3$ was chosen to maintain the perfect geometry symmetry of fractal structure. A diamond-shaped structure of dimensions 50 mm × 50 mm has been designed. With the scaling factor of 2.5, again a diamond-shaped structure with dimensions of 20 mm × 20 mm has been designed and iterated on the existing main diamond structure. This is the first iteration. During the second iteration, four diamond shapes of dimensions 8 mm × 8 mm have been designed and iterated on the main diamond structure. In the third iteration, eight more diamond shapes of dimensions 4 mm × 4 mm have been designed and iterated. The substrate used to design the proposed fractal antenna is FR4 of thickness $d = 1.6$ mm with relative permittivity $\varepsilon_r = 4.4$.

The dimensions of the Sierpinski fractal antenna can be approximately calculated by[15-23]

A. *Calculation of width (W):*
Width of the patch antenna is calculated by using

$$W = \frac{c}{2 f_0 \sqrt{(\varepsilon_r + 1/2)}}$$ (2.1)

where $c = 3 \times 10^8$ m/s.

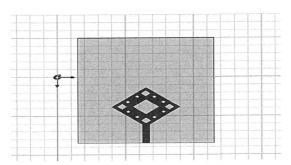

FIGURE 2.4 Geometry of proposed antenna.

B. *Calculation of actual length (L)*:

The effective length of patch antenna depends on the resonant frequency (f_0).

$$L_{\text{eff}} = \frac{c}{2f_0\sqrt{\varepsilon_{\text{reff}}}} \tag{2.2a}$$

where

$$\varepsilon_{\text{reff}} = \frac{\varepsilon_r + 1}{2} + \frac{\varepsilon_r - 1}{2}\left[1 + 12\frac{h}{W}\right]^{-1/2} \tag{2.2b}$$

Actual length and effective length of a patch antenna can be related as

$$L = L_{\text{eff}} - 2\Delta L \tag{2.3}$$

where ΔL is a function of effective dielectric constant $\varepsilon_{\text{reff}}$ and the width-to-height ratio (W/h)

$$\frac{\Delta L}{h} = 0.412\frac{\left(\varepsilon_{\text{reff}} + 0.3\right)\left((W/h) + 0.264\right)}{\left(\varepsilon_{\text{reff}} - 0.258\right)\left((W/h) + 0.8\right)} \tag{2.4}$$

C. *Calculation of feed width (W_f):*

To achieve 50-Ω characteristic impedance, the required feed width-to-height ratio (W_f/h) is computed as

$$\frac{W_f}{h} = \begin{cases} \dfrac{8e^A}{e^{2A} - 2} & \dfrac{W_0}{h} \leq 2 \\ \dfrac{2}{\pi}\left\{B - 1 - \ln(2B - 1) + \dfrac{\varepsilon_r - 1}{2\varepsilon_r}\left[\ln(B - 1) + 0.39 - \dfrac{0.61}{\varepsilon_r}\right]\right\} & \dfrac{W_0}{h} \geq 2 \end{cases} \tag{2.5a}$$

where

$$A = \frac{Z_0}{60} \sqrt{\frac{\varepsilon_r + 1}{2}} + \frac{\varepsilon_r + 1}{\varepsilon_r - 1} \left(0.23 + \frac{0.11}{\varepsilon_r} \right) \qquad (2.5b)$$

$$B = \frac{377\pi}{2Z_0 \sqrt{\varepsilon_r}} \qquad (2.5c)$$

D. *The number of iterations is*

$$N_n = 8^n \qquad (2.6)$$

E. *The ratio of fractal length is*

$$L_n = \left(\frac{1}{3} \right)^n \qquad (2.7)$$

F. *The ratio for the fractal area after the nth iteration is*

$$A_n = \left(\frac{8}{9} \right)^n \qquad (2.8)$$

where n is iteration nth stage number.

H. *The antenna is matched approximately at frequencies*

$$f_n \approx 0.26 \frac{c}{h} \delta^n \qquad (2.9)$$

where $\delta = 3$ is the log-period constant, n is a natural number, c is the speed of light in vacuum, and h is the height of the largest gasket.

2.4   RESULTS

2.4.1   DIMENSIONS OF SIERPINSKI DIAMOND FRACTAL ANTENNA

Two-element Sierpinski diamond antenna array with quarter-wave feed network is designed and fabricated with the design equations and is shown in Figures 2.5 and 2.6, respectively (Table 2.1).[2] Simulated and practical results such as reflection coefficient, voltage standing wave ratio (VSWR), and gain are observed in Figures 2.7–2.11, respectively (Table 2.2).[3]

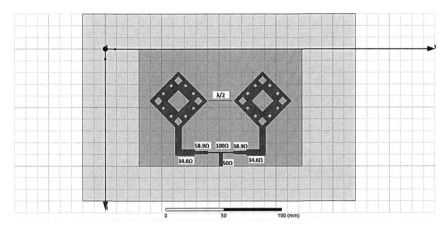

FIGURE 2.5 Geometry for two-element antenna array of quarter-wave feed network.

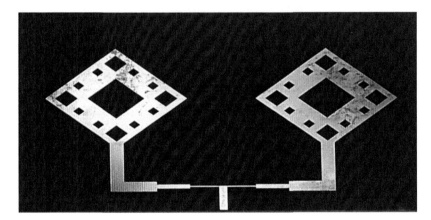

FIGURE 2.6 Fabricated patch of two-element antenna array of quarter-wave feed network.

TABLE 2.1 Dimensions of Single-element Sierpinski Diamond Fractal Antenna.

Parameter	Length (mm)
L1	35.355
L2	14.14
L3	5.656
L4	2.828
W1	5
A	24

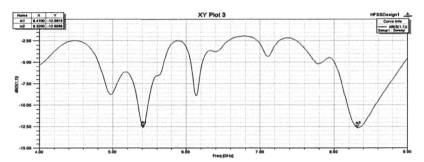

FIGURE 2.7   Reflection coefficient curve of the two-element antenna array of quarter-wave feed network.

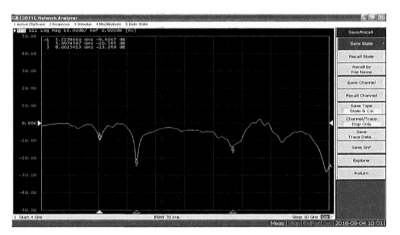

FIGURE 2.8   Reflection coefficient curve of fabricated two-element antenna array of quarter-wave feed network.

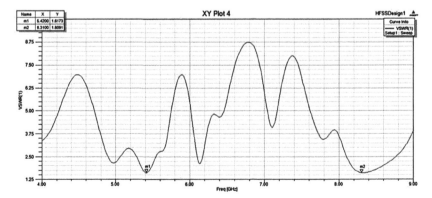

FIGURE 2.9   VSWR curve of the two-element antenna array of quarter-wave feed network.

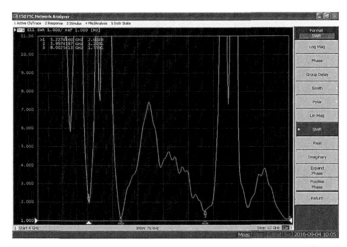

FIGURE 2.10   VSWR of fabricated two-element antenna array of quarter-wave feed network.

FIGURE 2.11   (See color insert.) Gain plot of the two-element antenna array of quarter-wave feed network.

TABLE 2.2   Results of Two-element Sierpinski Diamond Antenna Array with Quarter-wave Feed Network.

Resonant frequency (GHz)	Simulated	5.41
		8.3
	Measured	5.99
		8.00
Reflection coefficient (S11) (dB)	Simulated	−12.59 at 5.41 GHz
		−12.65 at 8.3 GHz
	Measured	−20.88 at 5.99 GHz
		−13.26 at 8.00 GHz

TABLE 2.2   *(Continued)*

VSWR	Simulated	1.61 at 5.42 GHz
		1.6 at 8.3 GHz
	Measured	1.2 at 5.99 GHz
		1.5 at 8.00 GHz
Gain (dB)	Simulated	5.2

2.5 CONCLUSION

In this chapter, a microstrip feed Sierpinski diamond fractal antenna array is designed and implemented by using quarter-wave feed network. Then, an improvement is observed in gain up to 5.2 dB (two elements). The linear/circular polarization behavior has also been achieved from the proposed structure. The proposed antenna can be used for various wireless communication applications. Also, it has been found that as iteration increases in fractal structure, number of bands also increases. The simulated and measured results are found to be in good agreement. It can be summed up that multibands are developed besides the resonance frequency. For further improvement in gain, we have to increase the number of elements in the array.

KEYWORDS

- **microstrip**
- **Sierpinski diamond**
- **quarter-wave transformer**
- **two-element antenna array**
- **WLAN**
- **fractal antennas**
- **T-split power divider**
- **mitered bend feed**

REFERENCES

1. Verma, S.; Ansari, J. A. Analysis of U-Slot Truncated Corner Rectangular Microstrip Patch Antenna for Broadband Operations. *Int. J. Electron. Commun.* **2015,** *69,* 1483–1488.

2. Gautam, A. K.; Bisht, A.; Kanaujia, B. K. A Wideband Antenna with Defected Ground Plane for WLAN/WiMAX Applications. *Int. J. Electron. Commun.* **2016,** *70,* 354–358.

3. Puente, C.; Borja, C.; Navarro, M.; Romeu, J. An Iterative Model for Fractal Antenna: Application to the Sierpinski Gasket Antenna. *IEEE Trans. Antennas Propag.* **2000,** *48,* 713–719.

4. Kingsley, N.; Anagnostou, D. E.; Tentzeris, M.; Papapolymerou, J. RF MEMS Sequentially Reconfigurable Sierpinski Antenna on a Flexible Organic Substrate with Novel DC-Biasing Technique. *IEEE Asia Pac. Microw. Conf.* **2008,** *16,* 1–4.

5. Choukiker, Y. K.; Behera, S. K. CPW-Fed Compact Multiband Sierpinski Triangle Antenna. *IEEE Antennas Wireless Propag. Lett.* **2010,** 1–3.

6. Liu, D.; Gaucher, B. Performance Analysis of Inverted-F and Slot Antennas for WLAN Applications. In *Proceedings of the 2003 IEEE AP-S International Symposium and USNC/URSI National Radio Science Meeting,* Vol. 2, Columbus, OH, June 23–27, 2003; pp 14–17.

7. Rowell, C. R.; Murch, R. D. A Capacitively Loaded PIFA for Compact Mobile Telephone Handsets. *IEEE Trans. Antennas Propag.* **1997,** *45,* 837–842.

8. Gianvittorio, J. P.; Rahmat-Samii, Y. Fractal Antenna: A Novel Antenna Miniaturization Technique and Applications. *IEEE Antennas Propag. Mag.* **2002,** *44* (1), 20–36.

9. Best, S. R. On the Significance of Self-similar Fractal Geometry in Determining the Multiband Behavior of the Sierpinski Gasket Antenna. *IEEE Antennas Wireless Propag. Lett.* **2002,** *1* (1), 22–25.

10. Mandelbrot, B. B. *The Fractal Geometry of Nature*; W.H. Freeman and Company: New York, 1983.

11. Puente, C.; Romeu, J.; Pous, R.; Ramis, J.; Hijazo, A. Small but Long Koch Fractal Monopole. *IEEE Electron. Lett.* **1998,** *34* (1), 9–10.

12. Borja, C.; Romeu, J. Multiband Sierpinski Fractal Patch Antenna. *Antennas Propag. Soc. Int. Symp., IEEE* **2000,** *3,* 1708–1711.

13. Harris, J. W.; Stocker, H. Scaling Invariance and Self-similarity and Construction of Self-similar Objects. *Handbook of Mathematics and Computational Science*; Springer-Verlag: New York, 1998; Sec. 4.11.1–4.11.2, p 113.

14. Puente-Baliarda, C.; Romeu, J.; Pous, R.; Cardama, A. On the Behavior of the Sierpinski Multiband Fractal Antenna. *IEEE Trans.* **1998,** *46* (4), 517–524.

15. Balanis, C. A. *Antenna Theory: Analysis and Design,* 3rd ed.; John Wiley & Sons, Inc.: Hoboken, NJ, 2005.

16. Rumsey, V. H. *Frequency Independent Antenna*; Academic Press: New York, London, 1966.

17. Cohen, N. L. Fractal Antennas Part 1. *Commun. Q.* **1995,** *Summer,* 5–23.

18. Hohlfeld, R. G.; Cohen, N. L. Self-similarity and the Geometric Requirements for Frequency Independence in Antennas. *Fractals* **1999,** *7,* 79–84.

19. Khan, S. N.; Hu, J.; Xiong, J.; He, S. Circular Fractal Monopole Antenna for Low VSWR UWB Applications. *Progress Electromagn. Res. Lett.* **2008,** *1,* 19–25.

20. Werner, D. H.; Ganguly, S. An Overview of Fractal Antenna Engineering Research. *IEEE Antennas Propag. Mag.* **2003,** *45* (1), 38–57.
21. Garg, R. *Micro Strip Antenna Design Handbook*; Artech House: Boston, MA, 2001.
22. Prabhakar, D.; Mallikarjuna Rao, P.; Satyanarayana, M. Design and Performance Analysis of Micro Strip Antenna Using Different Ground Plane Techniques for WLAN Application. *Int. J. Wireless Microwave Technol.* **2016,** *6* (4), 48–58.
23. Prabhakar, D.; Mallikarjuna Rao, P.; Satyanarayana, M. Characteristics of Patch Antenna with Notch Gap Variation for Wi-Fi Application. *Int. J. Appl. Eng. Res.* **2016,** *11* (8), 5741–5746.

CHAPTER 3

NOVEL CPW-FED TRIANGULAR-SHAPED ANTENNA FOR WIDEBAND APPLICATIONS

CH. SULAKSHANA[1*] and SWETHA RAVIKANTI[1,2]

[1]*Department of ECE, Vardhaman College of Engineering, Hyderabad, Telangana, India*

[2]*Department of ECE, National Institute of Technology, Warangal, Telangana, India*

Corresponding author. E-mail: sulakshana312ster@gmail.com

ABSTRACT

This paper presents a novel coplanar waveguide-fed semicircular patch on triangle-shaped antenna for wideband applications. The measured -10 dB impedance bandwidth is about 3.12 GHz (2.91–6.03 GHz) which is considered as wideband. The resonant peaks are observed at 3.45, 4.59, and 5.67 GHz. The first and third peaks fall in worldwide interoperability for microwave access (3.4–3.6 GHz) and wireless local area network (5.47GHz–5.725 GHz) applications. The effect of substrate dielectric constant and thickness have been evaluated. The results of antenna are simulated by using Zeeland's method of moments-based IE3D tool. Two-dimensional radiation patterns with elevation and azimuth angles, voltage standing wave ratio, return loss of -25.42dB, -27.78dB, and -34.38 dB at 3.45GHz, 4.59GHz, and 5.67 GHz, respectively, gain above 4 dB are obtained. The compact aperture area of the antenna is $32.2 \text{ mm}^2 \times 54.85 \text{ mm}^2$.

3.1 INTRODUCTION

Nowadays, wideband antennas are in great demand for wireless communication due to their lower radiation losses, wider bandwidth, lower dispersion, spectrum sensing, and radiation pattern stability.[1-3] With the recent wide and rapid development of wireless communications, there is a great demand in the design of low-profile, multiband antennas for mobile terminals.[4,5] Many antennas, such as the monopole and coplanar waveguide (CPW)-fed antennas with dual-band characteristics for wireless applications, have been reported.[6-9] Many bands for WLAN (wireless local area network) antennas are studied and published so far. The assigned bands according to IEEE 802.11 b/a/g are 2.4 GHz (2.4–2.484 GHz) and 5.2/5.8 GHz (5.15–5.35 GHz/5.725–5.825 GHz). The band 5.47–5.725 GHz is assigned for WLAN in Europe. The bands assigned for WiMAX (worldwide interoperability for microwave access) based on IEEE 802.16 are 2.5/3.5/5.5 GHz (2500–2690/3400–3690/5250–5850 MHz).[10-13]

Recently, due to its many attractive features such as wide bandwidth, low cross-polarization, radiation loss, less dispersion, uniplanar nature, no soldering point, and easy integration with active devices or monolithic microwave-integrated circuits, the CPW-fed antennas have been used as an alternative to conventional antennas for different wireless communication systems.[14]

In this chapter, a novel and simple antenna design has been carried out. A semicircular patch on triangle-shaped antenna fed by a CPW transmission line in a single-layer substrate is studied. With this CPW-fed scheme, the manufacture cost of the antenna can be reduced as low as possible. Details of the investigations based on simulations of the proposed antenna for wireless applications are described. First, a brief description of proposed antenna which includes antenna design and geometrical layout is presented in Section 3.2. The design methodology using IE3D is illustrated in Section 3.3 and the simulation results of return loss, radiation pattern, and antenna gain are given. Conclusion is drawn in Section 3.4. Simulations are carried out using Zeeland's "Method of Moments"-based commercial IE3D simulator.[15]

3.2   PROPOSED ANTENNA STRUCTURE AND DESIGN

The geometrical configuration of the proposed antenna is shown in Figure 3.1. The designed antenna is etched on a single layer of FR4 dielectric substrate whose dielectric constant is $\varepsilon_r = 4.4$ which is 32.2×54.85 mm^2 in dimension. The antenna is symmetrical with respect to the longitudinal direction, whose main structure is a triangular patch upon which semicircular patch is inscribed, with CPW feed line. The geometrical parameters are adjusted carefully and finally the antenna dimensions are obtained to be $L = 32.2$ mm,

$W = 54.85$ mm, $W_1 = 31$ mm, $L_1 = 28$ mm, $W_2 = 13$ mm, $L_2 = 2$ mm, $W_g = 5$ mm, $L_g = 8.1$ mm, $S = 3$ mm, $h = 5$ mm, $\varepsilon_r = 4.4$. The gap spacing between ground plane and CPW feed line is $g = 1.6$ mm. The substrate thickness is taken as $t = 1.6$ mm. The flare angle is $\theta = 60°$.

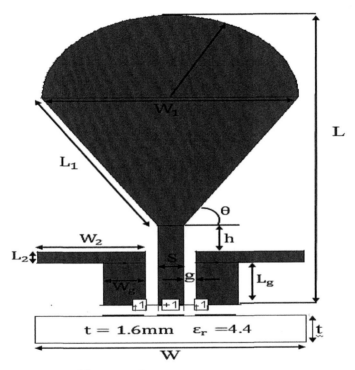

FIGURE 3.1   Layout of the proposed antenna.

The basis of the antenna structure is a patch element with a fixed length L_1 and a flare angle θ to lead the patch shape into either a strip monopole (i.e., $\theta = 90°$) with the same thickness as the signal strip or a triangular patch (i.e., $0 < \theta < 90°$) with varying width W_1 according to the fixed length L_1. Also noted that the patch is centered and connected at the end of the CPW feed line. In addition, the spacing between the patch and edge of the ground plane is h. The use of electromagnetic coupling effects of the ground planes to both the feed line and the patch element provides the wide impedance matching capability. Hence, the gap distance between the feed line and the ground planes mainly dominates the resonant mode of the upper band, while the flare angle (θ) is an important parameter to control the impedance bandwidth for this band. By properly selecting the antenna's geometric parameters numerically and experimentally, reduction of antenna size is also obtained.

3.3   INFERENCES FROM SIMULATED RESULTS AND DISCUSSIONS

To investigate the performance of the proposed antenna configurations in terms of achieving the required results, a commercially available moment method-based computer-aided design tool, IE3D, was used for required numerical analysis and obtaining the proper geometrical parameters in Figure 3.1.

3.3.1   RETURN LOSS

The simulated return loss coefficients are shown in Figure 3.2. It can be noted that for frequency range 2.91–6.03 GHz, a bandwidth of 3.12 GHz for a −10 dB return loss is observed with resonant peaks at 3.45, 4.59, and 5.67 GHz. The first and third peaks fall in WiMAX (3.4–3.6 GHz) and WLAN (5.47–5.725 GHz) applications. The return loss of −25.42, −27.78, and −34.38 dB at 3.45, 4.59, and 5.67 GHz, respectively, is observed. The return loss curve is obtained more accurately by taking 20 cells per wavelength. The voltage standing wave ratio (VSWR) values at 3.45, 4.59, and 5.67 GHz are observed to be 1.11, 1.08, and 1.04, respectively, which are closer to ideal value that is 1.

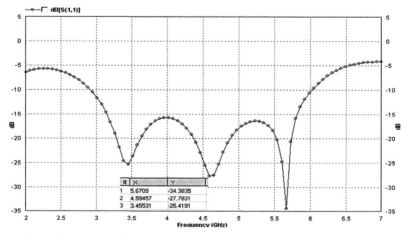

FIGURE 3.2   Return loss of the antenna.

3.3.2 RADIATION PATTERN WITH ELEVATION AND AZIMUTH ANGLES

The far-field radiation patterns at the operating frequency for the constructed prototype of the proposed antenna are also examined. Figures 3.3–3.8 depict, respectively, the simulated radiation patterns including E_θ and E_φ polarization patterns in the Azimuth cut (x–y plane) and the elevation cuts (y–z plane and x–z plane) for the antenna at the frequencies 3.45, 4.59, and 5.67 GHz, respectively.

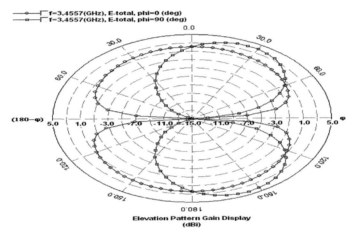

FIGURE 3.3   2D elevation pattern at 3.45 GHz.

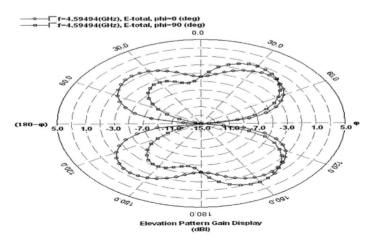

FIGURE 3.4   2D elevation pattern at 4.59 GHz.

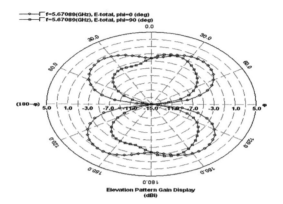

FIGURE 3.5   2D elevation pattern at 5.67 GHz.

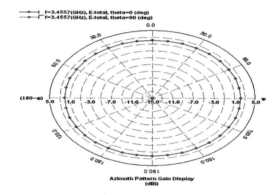

FIGURE 3.6   2D azimuth pattern at 3.45 GHz.

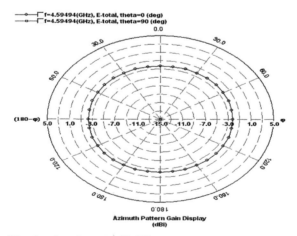

FIGURE 3.7   2D azimuth pattern at 4.59 GHz.

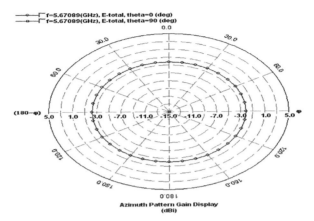

FIGURE 3.8 2D azimuth pattern at 5.67 GHz.

3.3.3 ANTENNA GAIN

The simulated antenna gain against frequency for the proposed antenna across the operating band is shown in Figure 3.9. It is observed that a gain of about 4.15, 4.5, and 2 dBi is observed at 3.45, 4.59, and 5.67 GHz, respectively. The antenna gain is above 2 dBi for the entire frequency range of 2.91–6.93 GHz.

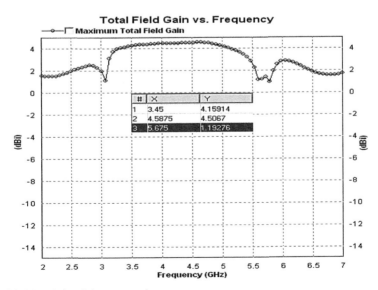

FIGURE 3.9 Gain of the proposed antenna.

3.3.4 PARAMETRIC STUDY BASED ON SUBSTRATE THICKNESS AND DIELECTRIC CONSTANT

The effect of substrate dielectric constant and thickness are evaluated by varying the dielectric constant of the substrate and thickness and the obtained return losses are compared. The compared resonant frequency, S_{11}, VSWR values, and bandwidths are tabulated in Tables 3.1 and 3.2, and corresponding curves are shown in Figures 3.10 and 3.11.

TABLE 3.1 Effect of Substrate Dielectric Constant on S_{11}, Bandwidth, and Resonant Frequency f_r.

ε_r	f_r (GHz)	S_{11} (dB) at f_r	Bandwidth (GHz)	VSWR
2.2	3.83	−14.38	3.96	1.47
	5.56	−20.63		1.20
	6.89	−33.52		1.06
3.38	3.63	−19.19	3.44	1.24
	5.06	−33.71		1.05
	6.18	−26.82		1.09
4.4	3.33	−25.75	3.04	1.18
	4.52	−29.41		1.08
	5.48	−32.26		1.05

TABLE 3.2 Effect of Substrate Thickness on S_{11}, Bandwidth, and Resonant Frequency f_r.

Thickness [t (mm)]	f_r (GHz)	S_{11} (dB) at f_r	Bandwidth (GHz)	VSWR
1.0	3.43	−25.27	3.13	1.12
	4.64	−29.71		1.07
	5.68	−28.16		1.09
1.5	3.45	−25.04	3.15	1.11
	4.68	−30.43		1.06
	5.73	−26.85		1.10
1.6	3.45	−25.41	3.11	1.11
	5.67	−34.38		1.08
	4.59	−27.79		1.04
2.0	3.33	−25.75	3.04	1.11
	4.52	−29.41		1.07
	5.48	−32.26		1.05

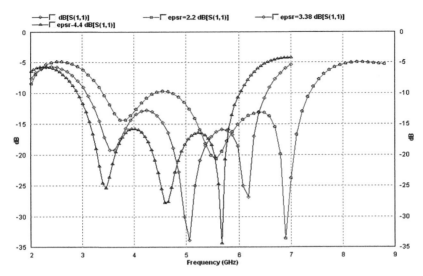

FIGURE 3.10 **(See color insert.)** Measured return loss for the proposed antenna with various dielectric constants.

Note: Other parameters are the same as in Figure 3.2.

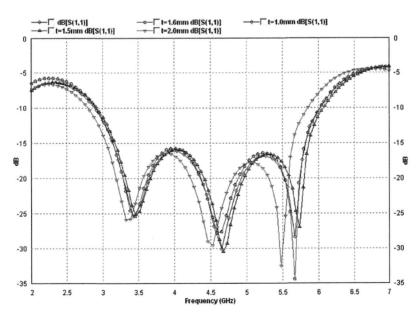

FIGURE 3.11 **(See color insert.)** Measured return loss for the proposed antenna with various substrate thicknesses.

Note: Other parameters are the same as in Figure 3.2.

After few repeated simulations using IE3D, it is found that for $\varepsilon_r = 4.4$ and $t = 1.6$ mm, the results are more accurate and resonant frequencies fall in WiMAX and WLAN applications. Moreover, it is observed that increasing the dielectric constant moves the resonant frequency to the left side of graph (decreases) and also decreases the bandwidth.

3.4 CONCLUSION

In this chapter, a CPW-fed semicircular patch on triangle-shaped antenna is proposed which does not require external matching circuit. The dimension of the antenna is 32.2×54.85 mm^2. By changing the substrate dielectric constant and thickness, the desired resonant frequency band is achieved. The resonant peaks at 3.45 and 5.67 GHz fall in WiMAX (3.4–3.6 GHz) and WLAN (5.47–5.725 GHz) applications. The desired antenna gain, radiation pattern, and VSWR (<2) are obtained. The results are found suitable for wideband wireless applications.

KEYWORDS

- **wideband antennas**
- **wireless communications**
- **triangular patch**
- **semicircular patch**
- **coplanar waveguide feed**
- **antenna gain**
- **return loss**

REFERENCES

1. Hossein, M.; Jam, S. Improved Radiation Performance of Low Profile Printed Slot Antenna Using Wideband Planar AMC Surface. *IEEE Antenna Wireless Propag. Lett.* **2016,** *11*, 1–12.
2. Rabah, M. H.; Setharandoo, D.; Addaci, R.; Bebbineau, M. Novel Miniature Extremely-Wide-Band Antenna with Stable Radiation Pattern for Spectrum Sensing Applications. *IEEE Antenna Wireless Propag. Lett.* **2015,** *14*, 1634–1637.

3. Sangaroon, O.; Anantrasirichai, N.; Rakluea, P.; Rakluea, C.; Poch, P. Study on CPW-Antenna for Wideband Coverage Mobile 4G/WLAN/WiMAX/UWB. In *7th International Conference on Information Technology and Electrical Engineering (ICITEE)*, Chiang Mai, Thailand, 2015.

4. Song, Y.; Jiao, Y.-C.; Zhao, G.; Zhang, F.-S. Multiband CPW-Fed Triangle-Shaped Monopole Antenna for Wireless Applications. *PIER* **2007,** *70,* 329–336.

5. Cui, Y.-Y.; Sun, Y.-Q.; Yang, H.-C.; Ruan, C.-L. A New Triple Band CPW Fed Monopole Antenna for WLAN and WiMAX Applications. *PIER M* **2008,** *2,* 141–151.

6. Liu, W.-C.; Hsu, C.-F. Dual-Band CPW-Fed Y-Shaped Monopole Antenna for PCS/WLAN Application. *Electron. Lett.* **2005,** *41* (18), 390–391.

7. Liu, W. C. Broadband Dual-Frequency Meandered CPW-Fed Monopole Antenna. *Electron. Lett.* **2004,** *40,* 1319–1320.

8. Li, J. Y.; Guo, J. L.; Gan, Y. B.; Liu, Q. Z. The Tri-band Performance of Sleeve Dipole Antenna. *J. Electromagn. Waves Appl.* **2005,** *19,* 2081–2092.

9. Shams, K. M. Z.; Ali, M.; Hwang, H. S. A Planar Inductively Coupled Bow-Tie Slot Antenna for WLAN Application. *J. Electromagn. Waves Appl.* **2006,** *20,* 861–871.

10. Nithisopa, J. K.; Nakasuwan, J.; Songthanapitak, N.; Anantrasirichai, N.; Wakabayashi, T. Design of CPW Fed Slot Antenna for Wideband Applications. *PIER* **2007,** *3* (7), 1124–1127.

11. Pan, C.-Y.; Horng, T.-S.; Chen, W.-S.; Huang, C.-H. Dual Wideband Printed Monopole Antenna for WLAN/WiMax Applications. *IEEE Antenna Wireless Propag. Lett.* **2007,** *6,* 149–151.

12. Tawk, Y.; Kabalan, K. Y.; El-Hajj, A.; Christodoulou, C. G.; Costantine, J. A Simple Multiband Printed Bowtie Antenna. *IEEE Antenna Wireless Propag. Lett.* **2008,** *7,* 557–560.

13. Wu, C.-M. Wideband Dual-Frequency CPW-Fed Triangular Monopole Antenna for DCS/WLAN Application. *Int. J. Electron. Commun. (AEU)* **2007,** *61,* 563–567.

14. Simons, R. N. *Coplanar Waveguide Circuits, Components and Systems*; John Wiley & Sons Inc.: New York, 2001.

15. IE3D Manual. *IE3D Manual Version, Release 12*; Zeland Software, Inc.: Fremont, CA, Oct. 2006.

CHAPTER 4

MULTIBAND FOUR PORT MIMO ANTENNA USING METAMATERIALS

F. B. SHIDDANAGOUDA[1*], R. M. VANI[2], P. V. HUNAGUND[3], and SIVA KUMARA SWAMY[1]

[1]*Sphoorthy Engineering College, Hyderabad 501510, Telangana, India*

[2]*University Science Instrument Center, Gulbarga University, Gulbarga 585106, Karnataka, India*

[3]*Department of Applied Electronics, Gulbarga University, Gulbarga 585106, Karnataka, India*

Corresponding author. E-mail: siddu.kgp09@gmail.com

ABSTRACT

A compact multiband four port multiple-input multiple-output (MIMO) antenna is presented using rectangular microstrip patch antenna. This work uses octagon complementary split-ring resonator (OCSRR) loaded on its ground plane. The designed MIMO antenna resonates at multiband frequency 2.36 Hz, 4.52 GHz, and 5.9 GHz, respectively, and for its isolation −22.09 dB, −24.22 dB, and −17.23 dB with overall bandwidth 344.5 MHz. The results of this designed MIMO antenna system shows a good isolation, bandwidth, peak gain, voltage standing wave ratio, and low envelop correlation coefficient which is promising for next generation wireless application systems.

4.1 INTRODUCTION

High data and low error rates are among the most demanding requirements of a general communication system. Multiple input multiple output (MIMO) is a technology that can provide both of these requirements. MIMO systems use multiple antennas for transmission and reception of the data. The data rate is linearly proportional to the number of antennas in the system over same bandwidth that not only increases the reliability but also uses the available bandwidth efficiently by increasing the overall capacity of the link.[1] Due to these properties, the fourth and fifth generation wireless systems will rely on MIMO. This chapter focuses designing such antenna for wireless communication terminals.

The microstrip patch antennas[2,11] are preferred because of their simple configuration like small size and simple structure. A patch antenna is a piece of metal on one side of a substrate of thickness h and dielectric constant ε_r and has a ground plane on the other side. These antennas have light weight, are compatible with other electrical device, and can be easily fabricated.[2] Hence, microstrip patch antennas[2,8,11] make them most suitable for wireless devices.

Split-ring resonators (SRR) are an integral component in realizing artificially engineered negative refractive index metamaterials.[1-3] Their electrical behavior is observed for an axial magnetic field for which these are seldom used in microstrip configuration. On the contrary, the complementary split-ring resonators (CSRR) are favorable for microstrip application as it responds to axial electric field. This property has been recently harnessed in designing planar microwave components.[4,5] Here in this chapter, the octagon complementary split-ring resonators (OCSRRs) have been realized on the ground plane of a rectangular microstrip antenna[9] and the corresponding behavior has been studied.

A four-element MIMO antenna with high isolation is presented in this chapter. The smallest edge-to-edge separation of the four symmetrical elements is λ/4 (where λ is the free space wavelength). To decrease the mutual coupling of the microstrip antenna arrays, a periodic metamaterial OCSRR structure as a spacer was applied in the ground plane. The proposed antenna can resonate at 2.36, 4.52, and 5.9 GHz and find application in several communication standard, and simulation results, including S-parameters; diversity gain, voltage standing wave ratio (VSWR) coefficient, and envelope correlation coefficients (ECCs) indicate that the proposed MIMO antenna can provide better diversity to increase data capacity of wireless communication systems.

This chapter is set up as follows. This section details the proposed four-element MIMO antenna layouts with metamaterial structure. The simulation result of the return loss, diversity gain, radiation pattern, VSWR coefficient, and ECCs are illustrated in Section 4.2 and Section 4.3 and the discussion is concluded with directions to future work in Section 4.4.

4.2 DESIGN PROCESS

4.2.1 CONCEPT AND CHARACTERIZATION OF THE OCTAGON SPLIT-RING RESONATORS

Recently, there has been a rising interest in metamaterial design; artificial negative magnetic permeability mediums are composed by SRR and introduced by Pendry.[12] Other examples of subwavelength metamaterial[7] structures are the broadside-coupled square, ring resonator (BC-SRR),[13] the capacitive-loaded loops,[14] and the spiral resonators,[15] which reduce the electrical size of the unit cell. In this chapter, we proposed a periodic OCSRR structure; a unit cell is depicted in Figure 4.1. As it can be seen from the figure, the metamaterial unit cell is composed of two nested split octagons that are etched on a dielectric substrate.

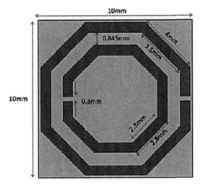

FIGURE 4.1 The unit cell structure.

The substrate is FR4 with dielectric constant equal to 4.4. Dielectric dimensions $L_s \times W_s$ are 10 mm × 10 mm and the thickness t is 1.6 mm. The strip width of each octagon is 0.6 mm. The sides of the octagons from the outer side to the inner side (S1, S2, S3, S4) are 4.0, 3.5, 2.8, and 2.3 mm, respectively. Both of the gaps, g, in the octagons are 0.3 mm and the distance between octagons, d, is 0.845 mm. To analyze this structure, it is embedded at the middle of a transverse electromagnetic (TEM) waveguide,[6] which has proper magnetic and electric boundary conditions on walls. The resonance frequency of this structure depends on the gap dimension (g). By increasing the gap, the capacitance in inductance–capacitance (LC) circuit model of the unit cell decreases. The decrement of the capacitance results in the increment of the resonance frequency of the structure.

4.2.2 GEOMETRY OF THE MIMO ANTENNA

The microstrip MIMO antenna as an initial design[10] consists of a FR4 substrate, a perfect conductor ground plane, and four patch elements. The distance of the elements is considered $\lambda/4$. The substrate's dimension and relative permittivity ε_r are 60 mm^3 × 60 mm^3 × 1.6 mm^3 and 4.4, respectively. The length and width of the patch are 11.35 mm^2 × 15.25 mm^2, length and width of quarter wave transformer are 4.90 mm^2 × 0.50 mm^2, length and width of 50 Ω feed line are 6.15 mm^2 × 3.05 mm^2, and the ground plane has a dimension of 60 mm^2 × 60 mm^2, respectively, shown in Figure 4.2.

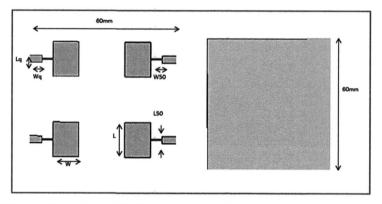

FIGURE 4.2   Conventional MIMO antenna front and bottom view.

After that, to more evaluate the performances of the proposed MIMO antenna, we have used four OCSRRs in ground plane as shown in Figure 4.3.

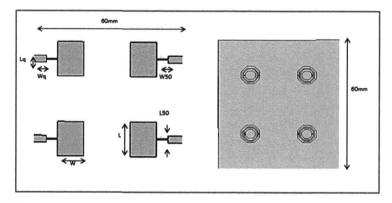

FIGURE 4.3   Octagon split-ring MIMO antenna front and bottom view.

4.3   RESULTS AND DISCUSSION

4.3.1   REFLECTION COEFFICIENT

The simulated reflection coefficient for the conventional MIMO antenna and designed MIMO antenna with OCSRRs is shown in Figures 4.4 and 4.5, respectively. The conventional MIMO antenna resonates at 5.90 GHz and for its isolation 22.15 dB with overall bandwidth 200 MHz, and MIMO antenna with OCSRRs can achieve better reflection coefficient rather than conventional MIMO antenna.

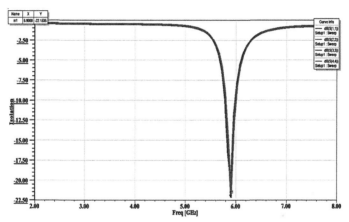

FIGURE 4.4   (See color insert.) Reflection coefficient of the conventional MIMO antenna.

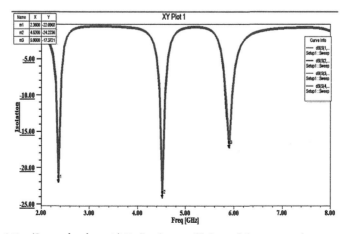

FIGURE 4.5   (See color insert.) Reflection coefficient of the proposed octagon split-ring MIMO antenna.

The designed antenna resonates at multiband frequency 2.36, 4.52, and 5.9 GHz, respectively, and for its isolation −22.09, −24.22, and −17.23 dB, with overall bandwidth 344.5 MHz, which can be considered good response for the designed antenna.

4.3.2 PEAK GAIN

The peak gain of the conventional microstrip MIMO patch antenna is 4.6 dBi and peak gain of octagon split-ring microstrip MIMO antenna is 5.21 dBi as depicted in Figures 4.6 and 4.7, respectively.

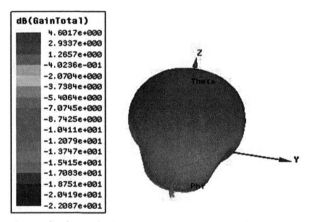

FIGURE 4.6   **(See color insert.)** 3D plot of conventional MIMO antenna gain.

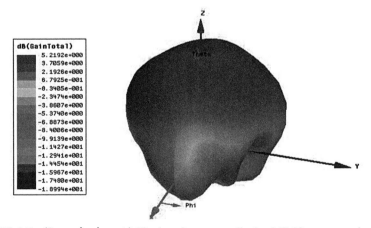

FIGURE 4.7   **(See color insert.)** 3D plot of octagon split-ring MIMO antenna gain.

4.3.3   VSWR

Normally in antenna application, the VSWR value is in between 1 and 2. Hence with MIMO antenna with OCSRRs, we get that the VSWR is 1.22, 1.13, and 1.31, respectively, at the resonance frequency of 2.36, 4.52, and 5.9 GHz. Hence, it is concluded that if VSWR is ≤ 2, then 89% of power transfers along the antenna, if VSWR < 1.7, then 93% of power transfer along the antenna, and if VSWR < 1.5, then 96% of power transfers. Therefore, our designed antenna VSWR value is less than 1.5; hence, good amount of power transfer along the antenna is shown in Figure 4.8.

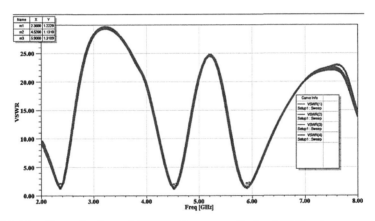

FIGURE 4.8   **(See color insert.)** VSWR of octagon split-ring MIMO antenna.

The ECC is the important parameter in MIMO system that can significantly influence the capacity of a MIMO system. This enveloped correlation coefficient between signals received by the antennas of the array can be computed through the S-parameters with the assumption that the incoming signals are uniformly distributed, that is, the direction of arrival of each multipath component has equal probability. The correlation coefficient between any two elements of the array is given as

$$\rho_{i,j} = \frac{\left| s_{ii}^* s_{ij} + s_{ji}^* s_{jj} \right|}{\left(1 - |s_{ii}|^2 - |s_{ji}|^2\right)\left(1 - |s_{jj}|^2 - |s_{ij}|^2\right)}$$

From Figure 4.9, it is clear that the ECC is very low over the whole band which indicates very good isolation between the four antennas. The maximum ECC is less than 0.03.

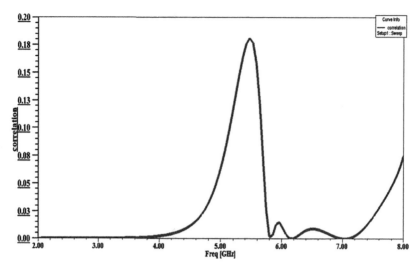

FIGURE 4.9   **(See color insert.)** Correlation coefficient of octagon split-ring MIMO antenna.

4.4   CONCLUSION

A compact multiband four-port MIMO antenna with improved isolation multiband is proposed and designed. A good isolation multiband is achieved by introducing octagon split ring loaded on its ground plane of microstrip patch antenna. The designed antenna has small size with simple geometry which makes it highly suitable for integration into system circuits. The characteristics obtained show that the proposed MIMO antenna is fit for wireless communication applications.

KEYWORDS

- MIMO
- communication system
- wireless systems
- microstrip patch antennas
- metamaterials
- octagon complementary split-ring resonators

REFERENCES

1. Balanis, C. A. *Antenna Theory: Analysis and Design*, 2nd ed.; John Wiley & Sons: Hoboken, NJ, 1982 (Chapter 14).
2. Huang, J. *A Review of Antenna Miniaturization Techniques for Wireless Applications*; Jet Propulsion Laboratory, California Institute of Technology: Pasadena, CA, 2001.
3. Waterhouse, R. B.; Targonski, S. D.; Kokotoff, D. M. Design and Performance of Small Printed Antennas. *IEEE Trans. Antennas* **1998**, *46* (11), 1629–1633.
4. Trippe, A.; Bhattacharya, S.; Papapolymerou, J. Compact Microstrip Antennas on a High Relative Dielectric Constant Substrate at 60 GHz. In *IEEE Antennas and Propagation (APSURSI)*, Spokane, July 2011; pp 519–520.
5. Engheta, N. An Idea for Thin Sub-wavelength Cavity Resonator Using Metamaterials with Negative Permittivity and Permeability. *IEEE Antennas Wireless Propag. Lett.* **2002**, *1* (1), 10–13.
6. Alu, A.; Engheta, N. Guided Modes in a Waveguide Filled with a Pair of Single-negative (SNG), Double-negative (DNG), and/or Double–Positive (DPS) Metamaterial Layers. *IEEE Trans. Microwave Theory Tech.* **2004**, *MTT-52* (1), 199–210.
7. Alu, A.; Bilotti, F.; Engheta, N.; Vegni, L. Sub-wavelength, Compact, Resonant Patch Antennas Loaded with Metamaterials. *IEEE Trans. Antennas Propag.* **2007**, *55* (1), 13–25.
8. Lee, Y.; Hao, Y. Characterization of Microstrip Patch Antennas on Metamaterial Substrates Loaded with Complementary Split-ring Resonators. *Microw. Optic. Technol. Lett.* **2008**, *50* (8), 2131–2135.
9. Bazrkar, A.; Gudarzi, A.; Mahzoon, M. In *Miniaturization of Rectangular Patch Antenna Partially Loaded with U-Negative Metamaterials*. International Conference on Electronics, Biomedical and Its Applications (ICEBEA), Dubai, Jan 2012; pp 289–292.
10. Ye, J.; Cao, Q.; Tam, W. In *Design and Analysis of a Miniature Metamaterial Microstrip Patch Antenna*. IEEE Antenna Technology (iWAT), Hong Kong, March 2011; pp 290–293.
11. Pattnaik, Sh. S.; Joshi, J. G.; Devi, S.; Lohokare, M. R. In *Electrically Small Rectangular Microstrip Patch Antenna Loaded with Metamaterial*. IEEE Antennas Propagation and EM Theory (ISAPE), Guangzhou, Nov 29, 2010–Dec 2, 2010; pp 247–250.
12. Kamtongdee, C.; Wongkasem, N.; Charoen, B.; Matra, K. Development of Compact Microstrip Antennas Using Metamaterials. *ITC-CSCC* **2009**, *7*, 290–293.
13. Hou, D.; Xiao, S.; Wang, B.-Z.; Jiang, L.; Wang, J.; Wei, H. Elimination of Scan Blindness with Compact Defected Ground Structures in Microstrip Phased Array. *IET Microwave Antennas Propag.* **2009**, *3* (2), 269–275.
14. Keowsawat, P.; Phongcharoenpanich, C.; Kosulvit, S. Mutual Information of MIMO System in a Corridor Environment Based on Double Directional Channel Measurement. *J. Electr. Waves Appl.* **2009**, *23*, 1221–1233.
15. Liu, Y.; Zhao, X. Enhanced Patch Antenna Performances Using Dendritic Structure Metamaterials. *Microwave Opt. Technol. Lett.* **2009**, *51*, 1732–1738.

CHAPTER 5

E-SHAPE TOP-LOADED OCTAGONAL PATCH ANTENNA FOR SMALL-FREQUENCY APPLICATIONS

P. VENU MADHAV[1*] and M. SIVAGANGA PRASAD[2]

[1]*Department of ECE, KL University, Guntur, Andhra Pradesh, India*

[2]*Department of Electronics and Communication, KKR and KSR Institute of Technology and Sciences, Guntur, Andhra Pradesh, India*

Corresponding author. E-mail: Venu7485@gmail.com

ABSTRACT

A compact coplanar octagonal microstrip patch antenna operating at a frequency range of 1.2–10.0 GHz is presented and discussed in this paper. The circle in the center of the patch and the L-shaped stubs that are enclosed inside the inverted E-shape act as a notching element in attaining the frequency of operation. The patch is modified further to have dual band characteristics; fine tuning the impedance match makes it suitable for wireless, mobile, and C-band applications. The antenna is modeled by means of FR4_Epoxy material as substrate with a thickness of 1.6 mil. The design of the antenna is simulated using high-frequency structure simulation. To operate in multiple bands of 1.1–2.1 GHz, 2.8–3.4GHz, and 6.1–6.9 GHz, the antenna has been elevated. For optimizing the size of the octagonal area, the particle swarm optimization method is also considered for analyzing. The parametric studies on the antenna design after fabrication of prototype have been presented. This includes measurement of voltage standing wave ratio, radiation pattern, peak gain, etc.

5.1   INTRODUCTION

Short-range wireless communication systems started playing a key role in the design of any commercial application. It has advantages like high data rate, low power consumption, and wider bandwidth. Due to current trend in communication system, many researchers are trying to design and develop cost-effective, capable of operating at multiband frequencies, antennas which can be easily integrated with radio frequency (RF) wave circuits. The most important limitation of a microstrip patch antenna[1] is its narrow band operation. Although some investigations are carried out to enhance the bandwidth limitation, still it is an addressable problem. In this chapter, we propose a simple coplanar antenna designed for operating at a frequency range of 2–10 GHz and attained multiple resonant frequencies at 2.3, 2.9, and 6.3 GHz (nearly), thus making it suitable application for operating in the wireless LAN. The design of the antenna is modeled using HFSS (high-frequency structure simulation) software and the result is presented and discussed.

5.2   ANTENNA STRUCTURE

5.2.1   RESONANT FREQUENCY (f_r)

The fundamental design of the antenna structure depends on the operating frequency which in turn depends on the specific application. Selection of operating frequency is 2.5 GHz as it lies within the mobile communication systems operation range (2–5 GHz). The selection of a substrate substantial is also equally significant.

5.2.2   SELECTION OF A SUBSTRATE MATERIAL (ε_r)

The dielectric substantial carefully chosen for the proposed model is FR4_ Epoxy; the use of dielectric substrate enhances the ability to maintain excellent mechanical, physical, and electrical properties at elevated temperatures and also ensures safety and consistency. These are used to ease the size of the antenna due to complex permittivity and can help to produce displacement currents which results in producing time-varying electric fields and creates propagating electromagnetic field. A substrate also enhances the radiating capability of the antenna.

The FR4_Epoxy has the following notable features (Table 5.1).

TABLE 5.1 Properties of FR4_Epoxy.

Parameters	Values
Dielectric constant	4.36
Loss tangent	0.013
Water absorption	<0.25%
Tensile strength	<310 MPa
Volume resistivity	8 10^7 MΩ cm
Surface resistivity	2×10^5 MΩ
Breakdown voltage	55 kV
Peel strength	9 N/mm
Density	1850 kg/m^3

5.2.3 HEIGHT OF THE DIELECTRIC MATERIAL (H)

The height of the dielectric substantial is to be chosen in such a way that it should not affect the ground wave propagation. Hence, the elevation of the dielectric substrate is taken as 1.64 mm.

The main design factors of the patch are as follows:

- $f_r = 2.2$ GHz
- $\varepsilon_r = 4.3$
- $h = 1.64$ mm

5.3 DEVELOPMENT PROCESS

The optimization process of the microstrip antenna parameters is to be done at the initial stage using particle swarm optimization (PSO)[2,3] for predetermined shapes of the antenna and helps to levy conditions during optimization method of the antenna shape and support in fabrication. The design style does not boundary the shape of the antenna.

Calculate the resonant frequencies of a polygonal patch antenna, along with other parameters.

$$f_r = \frac{X_{nm}C}{16r\sin(\theta/2)\sqrt{\varepsilon_r}} \qquad (5.1)$$

where c is the velocity of light.

$$X_{nm} = K_{nm} r$$
$$\theta = 45°$$

$$k = \frac{2\pi\sqrt{\varepsilon_r}}{\lambda_0} \qquad (5.2)$$

By solving Equation (5.1), the resonant frequency for the first TM_{nm} modes of an octagonal patch antenna can be achieved.

$$f_r = \frac{90.24}{r\sqrt{\varepsilon_r}} \qquad (5.3)$$

where f_r is the resonant frequency, r is the radius in mm, and ε_r is the dielectric permittivity.

5.3.1 FEEDING STRUCTURE

A coplanar waveguide (CPW) antenna requires only one single conducting layer simplifying the fabrication process and also avoids alignment problems. A CPW loop is shown to be an effective low VSWR feed for microstrip antenna; besides, it also offers features like less radiation loss, less dispersion, easy integration with monolithic MMIC, and effective control characteristic impedance.

5.3.2 BANDWIDTH

A straightforward method for improving bandwidth is to increase the thickness of the substrate, but it will increase the surface-wave power and thus effect the radiated power to decrease. There are abundant methods for enhancing bandwidth of antenna like use of manifold resonances, using folded patch feed, etc.

5.3.3 OPTIMIZATION

During this study, to curtail the size of the octagon, PSO has been adopted with constraints on frequency of the fundamental mode. The significance of the PSO method is to optimize different natural variables which can be further used to minimize physical dimension, thickness of the substrate, and other factors of the antenna. In this chapter, the approach is to optimize the patch dimension.

5.3.4 PSO APPROACH

The PSO practice is an evolutionary calculation technique and varies from genetic algorithm. In PSO, a bird flock's behavior simulates the population dynamics, where distribution of data takes place and every entity can profit from the previous experience or discoveries. In this context, each particle is preserved as a point in a dimensional space, defined as a moving point in space.

In a PSO algorithm,[5] evolutionary operators such as crossover to over-write, mutation for k variables optimization,[4,6] and a flock of bits are placed into the k-dimensional space with randomly chosen positions knowing their L-best values and the position. In the k dimension, the position of each particle is accustomed according to its ability to transmit or radiate. For example, the kth particle is symbolized as mi $= $ mi$_1$, mi$_2$, mi$_3$, ... mi$_n$ in the dimensional space and the best previous position of the kth particle is documented as mbesti $=$ mbesti$_1$, mbesti$_2$, mbesti$_3$, ... mbesti$_n$, and the index among the best particle in the particle group is represented as (s)-best.

5.3.5 OPTIMIZATION OF OCTAGONAL PATCH

For an octagonal patch, the angle and the radius of the octagon are the significant parameters that are to be optimized for a resonant frequency of 1–2 GHz.

The resonant frequency of the TMnm mode of the octagonal patch from eq (5.3) is

$$f_r = \frac{90.24}{r\sqrt{\varepsilon_r}}$$

where r is the radius of the octagonal shape and ε_r effective is given as

$$\varepsilon_r = \frac{\varepsilon_r + 1}{2} + \frac{\varepsilon_r - 1}{2}\left(1 + \frac{12h}{a}\right)^{-1/2}$$

The second frequency resonance is attributed to the patch and it is seen that with an increase in the patch height, there is a resonance shift toward the lower side.

5.3.6   SMITH CHART

A Smith chart is a simple depiction of all possible composite impedances with respect to reflection coefficient. By definition, reflection coefficient is a circle of radius 1 in a multifaceted level.

The main purpose of Smith chart illustration is to recognize all possible impedances. The normalized impedance is represented in a Smith chart by using curls that categorize the normalized resistance r (real part) and normalized reactance x (unreal).

$$z(d) = \text{Re}(z) + j\,\text{Im}(z) = r + jx \tag{5.4}$$

Hence, the representation of reflection coefficient in expressions of its matches is

$$\tau(d) = \text{Re}(\tau) + j\,\text{Im}(\tau) \tag{5.5}$$

The Smith chart is a polar plot of complex reflection coefficient and is mathematically defined as a 1-port scattering parameters (S_{11}).

The reflection coefficient is used to characterize a load which may be admittance, gain, and *trans*-conductance and is useful for RF frequencies.

Notable points from a Smith chart.

- All circles intersect at the coordinate (1,0)
- The zero Ω circle is the largest one and there is no resistance
- Immeasurable resistance is reduced to one point at (1,0)
- Choosing resistance value can be simple just by altering alternative circle conforming to new value

5.3.7   SIMULATION MODEL

Structural analysis: An octagon is a closed loop with equal sides and internal angle of 135°. It is necessary to calculate the radius of the octagon to be 22 mm and designed for the resonant frequency of 1.9 GHz. The size of the octagon is considered as $2a^2(1+\sqrt{2})$, where a is the side of the octagon. The antenna is fed by a microstrip line with a dimension of 3 mm in width.

The design of the proposed antenna is developed stage-wise to check the accuracy of the model. In the first stage, a simple octagon shape with 1.5 cm diameter is designed on a coplanar plane using microstrip line feed. The ground plane is chosen such that it is larger by two times to that of the radiating patch. The resonating frequency obtained is above 10 GHz (Fig. 5.1).[10] In the second stage, a circle with a radius of 0.7 mm (nearly) has been used as a notch to bring down the frequency of operation to less than 10 GHz (Fig. 5.2). In stage 3, a top-loaded inverted E has been introduced to attain a resonating frequency less than 5 GHz (Fig. 5.3). In stage 4, two mirror L-stubs[7] are used to make the antenna resonate at multiple frequencies, thus making it suitable for C-band, wireless, and mobile applications (Fig. 5.4).

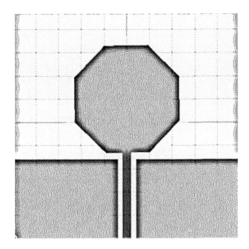

FIGURE 5.1   Coplanar octagonal antenna. Adapted from Ref. [10].

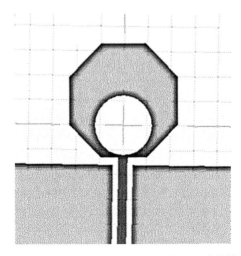

FIGURE 5.2   Circle via in octagon (stage 2). Adapted from Ref. [10].

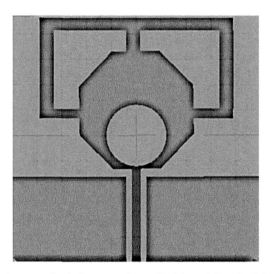

FIGURE 5.3   E-shape top-loaded octagon (stage 3). Adapted from Ref. [10].

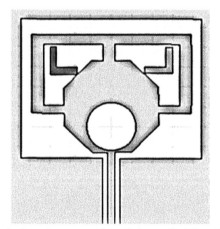

FIGURE 5.4   L-shaped wings for octagonal patch (stage 4). Adapted from Ref. [10].

The imitation tool is used in evaluating the performance of the antenna which is established on the method of moment's technique and used for computing VSWR, return loss, and gain of the suggested antenna. The return loss specifies the volume of power that is lost to load and does not return as reflection. Figure 5.5 shows that during the first stage, the antenna shows a return loss above 10 GHz and later after stage 4, the antenna is able to be operated in multiple bands ranging from 1.9 to 2.4 GHz, 2.9 to 3.2 GHz, and 5.3 to 6 GHz. Table 5.1 shows that as the stage increases, the fundamental frequency is shifted to the lower side and is suitable for multiband applications. The patch-covering geometries were inspected ideally, and reasonable values of resonant frequency, return loss, and gain are inspected and related. The comparative table shows that final patch which gives decent results in contrast with stage progress. It is observed that the antenna has minimum VSWR, progress in reflection coefficient, three frequency bands with considerable bandwidth, and gain.

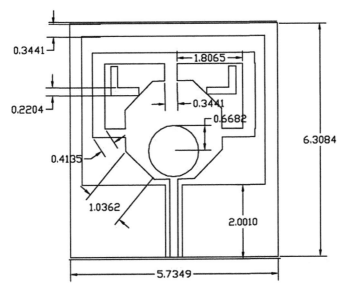

FIGURE 5.5 Dimensions of the proposed antenna model. Adapted from Ref. [10].

5.3.8 FABRICATION

The patch is made up on a low-cost FR4_Epoxy substrate with a material thickness of $h = 1.66$ mm and permittivity ε_r of 4.36. To get better correctness, the antenna is sketched using AutoCAD and is fabricated using photolithographic process. The overall length and width of the antenna are fitted into a 5 cm × 5 cm dimension. The patch and ground are disjointed by a closed cell and it aids to obtain wider bandwidth and higher gain. During the initial iteration, the fundamental resonating frequency of the antenna is above 15 GHz but in the second reiteration when an inverted E is placed on the top of the main path, a shift in the resonant frequency has been observed to the lower side. In the third and fourth iterations when two stubs are added to the main patch that lies internal to the inverted E shape, there is a significant shift of the resonant frequency between 1 and 2 GHz. The rectangular plot of Figure 5.6 represents that when dual L-stubs are added to the main patch, additional resonance also occurred at 3, 6.5, and 9.5 GHz.

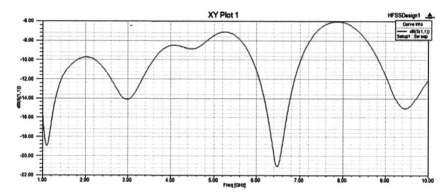

FIGURE 5.6   Rectangular plot of S_{11}.

The increase in the slot width results in increase in impedance bandwidth. The impedance corresponding curve moves toward the inductive area of the Smith chart and is as shown in Figures 5.7–5.9.

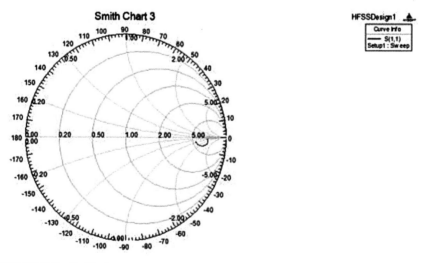

FIGURE 5.7   Smith chart.

The maximum gain obtained by the antenna is at 1.2 GHz and between 6 and 7 GHz. The variation of gain at other resonating frequencies of the antenna is tabulated below.

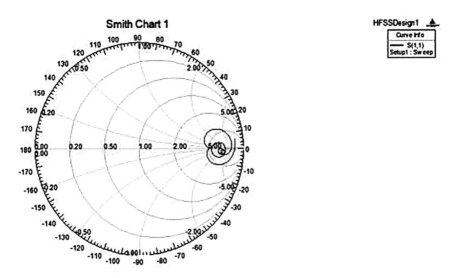

FIGURE 5.8   Inductive impedance shift.

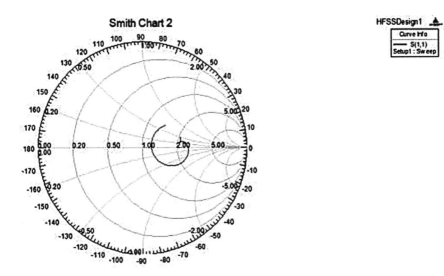

FIGURE 5.9   Smith chart.

5.4 CONCLUSION AND DISCUSSION

The antenna has been computer-generated using HFSS, the sphere-shaped scanning system was utilized for near field antenna measurement using standard spherical wave expansion techniques, and the radiation of the antenna can be fully described by a set of prototypical coefficients.

The inverted E-shaped radiation faces of the antenna are also studied and the measured radiation outlines[8] for Azimuth and elevation are as shown in the figure. The designed antenna displays a good broadband radiation pattern. The peak cross-polarization level of the antenna is observed about −8 to −22 dB. It is prominent that the radiation characteristics of the recommended patch are better to the predictable patch antenna. The parametric studies also address the effect of width[9] and length of the patch, height, and air gap on the performance of the antenna.

ACKNOWLEDGMENTS

I would like to thank Dr. Sivaganga Prasad Garu, for sharing his pearls of wisdom and providing awareness and greatly assisting the research, an unceasing support of my PhD study and allied research for his patience, motivation, and immense knowledge. His constant supervision aided me all the time.

KEYWORDS

- wireless communication systems
- high data rate
- low power consumption
- wider bandwidth
- microstrip patch antenna
- dielectric substrate
- reflection coefficient

REFERENCES

1. Ang, B. K.; Chung, B. K. A Wideband E-Shaped Microstrip Patch Antenna for 5–6 GHz Wireless Communications. *PIER* **2007,** *75,* 397–407.
2. Hussein, M. K. Design of Microstrip Antenna Using Particle Swarm Optimization. *Wasit J. Sci. Med.* **2011,** *4* (2), 162–173.
3. Ali Jawad, K. A New Compact Size Microstrip Patch Antenna with Irregular Slots for Handheld GPS Applications. *Eng. Technol. J.* **2008,** *26* (10), 1241–1246.
4. Singh, J. M.; Mishra, M.; Sharma, P. Design and Optimization of Microstrip Patch Antenna. *IJETTCS* **2013,** *2* (5), 139–141.
5. Ali, M.; Dougal, R.; Yang, G.; Hwang, H. S. Wideband (5–6-GHz WLAN Band) Circularly Polarized Patch Antenna for Wireless Power Sensors. *IEEE Antennas Propag. Soc. Int. Symp. Dig.* **2003,** *2,* 34–37.
6. Minasian, A. A.; Bird, T. S. Complimentary Particle Swarm Antennas for Next Generation Wireless Communication Systems. In *Proc. ISWCS,* Paris, France, Aug. 2012; pp 895–898.
7. Singh, A. K.; Meshram, M. K. Slot Loaded Short Patch for Dual Band Operation. *Microwave Opt. Technol. Lett.* **2008,** *50* (4), 1010–1017.
8. Yorozu, Y.; Hirano, M.; Oka, K.; Tagawa, Y. Electron Spectroscopy Studies on Magneto-optical Media and Plastic Substrate Interface. *IEEE Transl. J. Magn. Jpn.* **1987,** *2,* 740–741 [*Digests 9th Annual Conf. Magn.* Japan, p. 301, 1982].
9. Wong, K. L. *Planar Antennas for Wireless Communications*; John Wiley & Sons, Inc.: Hoboken, NJ, 2003.
10. Hariyadi, T. In *A Coplanar Waveguide (CPW) Wideband Octagonal Microstrip Antenna*; Conference: Information and Communication Technology (ICoICT), 2013.

CHAPTER 6

A SURVEY ON MINIATURIZATION OF CIRCULARLY POLARIZED ANTENNAS FOR FUTURE WIRELESS COMMUNICATIONS

SWETHA RAVIKANTI* and L. ANJANEYULU

Department of ECE, National Institute of Technology, Warangal, Telangana, India

Corresponding author. E-mail: swetha.rks@nitw.ac.in

ABSTRACT

Circular polarized microstrip patch antennas are more in demand for next generation with compact size. A survey on various techniques like sequential feed, folded patch, substrate-integrated metallic wall structure, slits in miniaturization of a patch antenna, and various parameters like axial ratio, beam width, gain, and size reduction in respect of meeting the required levels for future wireless communication is considered.

6.1 INTRODUCTION

In wireless communication systems, circularly polarized microstrip antennas (CPMAs) grasped much attention in this new era. When the systems are in rotating motion with respect to orientation, circular polarization is useful in such cases. In circular polarization, the transmitter and receiver with respect to orientation are considered; independent data transmission is allowed. Compact CPMAs are used in applications such as handheld radio frequency identification (RFID) reader and portable wireless devices. In CPMAs, single and dual feed structures are used.[1] As far as feed location is considered with the single-feed configuration at the radiator, structure is excited in orthogonal mode at 90° phase shift for circular polarization (CP). In microstrip antennas,[4–11]circular polarization provides large bandwidth in dual feed structures compared to the single feed.[2,3] But in the dual feed configuration, large size is required in ground plane for feeding network. These perturbation techniques are most popular in size reduction of CPMAs. Truncating a pair of square patch corners in square microstrip antenna which is symmetric is a well-known method of producing CP radiation.[5] The size reduction can be done by a technique called symmetric slits in microstrip patches which has been proposed by Chen et al.[9] CP radiation can be achieved using the technique conventional symmetric corner truncating and can be designed with slotted ground plane.[10] Truncated corner method for circular polarization which has no reduction in size is considered. The size reduction is also possible with substrate-integrated metallic wall structure. Along the diagonal direction of a patch antenna are used symmetric slits and tails for smaller designs in CPMA.[11] However, Nasimuddin et al.[12] have proposed the study of the compact CPMAs with asymmetric slit patch radiator. The above designs had CP characteristics which achieve less than 4% only in axial ratio (AR) bandwidth and for navigation purpose, multiple systems cannot be accommodated at the same time with lesser axial bandwidth ratio. Techniques like wideband feeding mechanism[12–14] and sequentially phase feeding networks[15–19] are used to increase the AR up to 10% and have been employed. However, the size of the printed circuit or the ground plane is enlarged which resulted in complicated geometries as was the same case with the other above wideband feeding techniques. Sequential feeding technique is proposed by Hau et al.[20] to obtain wide AR bandwidth. Another technique called folding the patch is used to improve beam width, impedance band-width, and reduction in resonant frequency, but with a drawback of large size and design complexity[21]. With the above techniques, the radiators are larger in size and to reduce it, other techniques such as folding, cutting the slots, and adding tails are less effectively used when compared to loading high dielectric substrate, using shorting pins/walls. Shorting walls or pins is the technique proposed by Wong et al.[24] in size reduction for the patch antenna.

6.2   TECHNIQUES

6.2.1   SEQUENTIAL FEEDING

Sequential feeding mechanisms for patch radiator with circular polariza-tion are analyzed. The antenna has two main portions, a double-sided printed-circuit-board (PCB) at the bottom and a metal patch at the top which has a shape of square. In a ground plane, four Γ-shaped slots are etched on the top side of the double-sided PCB and have a microstrip line at the bottom side of the double-sided PCB. It is observed that centers of the double-sided PCB and the metal patch are aligned together but they are located in z-direction at different level. They are separated by a foam-supporting block with a thickness h1, which is 11 mm. The PCB has a thickness (h2) of 1 mm, length of 60 mm, and dielectric constant ϵ_2 = 2.65. The square patch has a length (L) of 43.4 mm and a thickness of 0.3 mm as shown in Figures 6.1–6.3. This antenna has stable radiation pattern, wide impedance bandwidth, and wide AR bandwidth. The measured standing wave ratio) (<1.5) and 3 dB AR bandwidth of the antenna are over 16.5% and 13.3%, respectively. The peak gain of the antenna is 7.4 dBic at 2.55 GHz.[20,22]

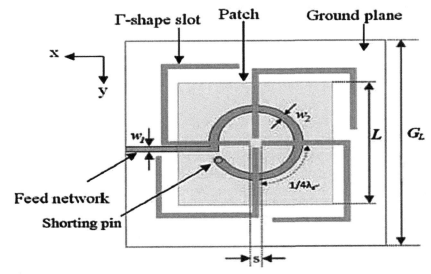

FIGURE 6.1 Top view.

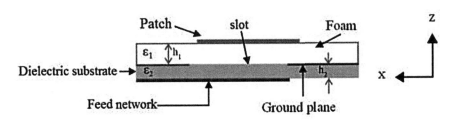

FIGURE 6.2 Side view.

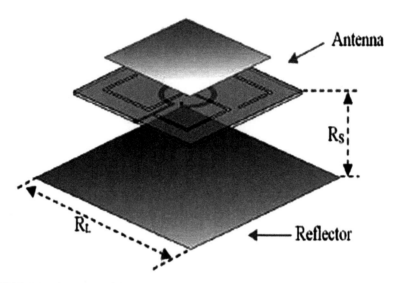

FIGURE 6.3   Geometry of the antenna with reflector.

6.2.2   SUBSTRATE-INTEGRATED METALLIC WALL STRUCTURE

A circularly polarized patch radiator using substrate-integrated metallic wall is shown in Figure 6.4. It is constructed with two layers of substrates that are Substrate1 and Substrate2 with thickness H_1 and H_2. The metallic walls are integrated in Substrate1 and the radiating patch is printed on Substrate2 which are stacked together. For a patch radiator underneath the ground, plane is fed through a coaxial probe which is connected to SMA connector. Each corner of radiating patch consists of four groups of metallic walls. In each metallic wall, substrate is integrated by vertical vias which is realized by connecting. The pin diameter of flat strips has the same width. Inductance is created at the vertical portion of the walls and along the diagonal of the patch; the distance between two groups of metallic walls is characteristic by c. d indicates the separation between two parallel walls is considered. At the meantime, the antenna resonates at 3.24 GHz and the separation between two parallel walls provides the capacitance that has obtained AR of 2.16% with 80% of reduction in size when compared with conventional half-wavelength square patch antenna. From this technique, the size of the antenna is reduced with the help of inductive and capacitive loading effect.[21]

but only depends on area of the slits. Microstrip patch radiators of square shape can be realized for circular polarization using a single coaxial feed structure for a compact antenna size; asymmetrical slits are cut along the diagonal direction. The performances of the proposed antennas with several asymmetric slit shapes onto the patch radiators are compared. The measured 10-dB return loss and 3-dB axial-ratio bandwidths of the antenna prototype are around 2.5% and 0.5%, respectively.[12]

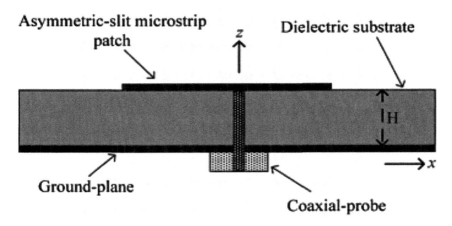

FIGURE 6.6 Cross-sectional view of asymmetric slit compact CPMA.

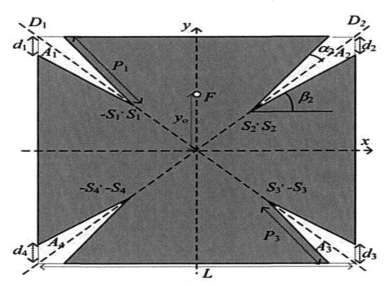

FIGURE 6.7 Top view of asymmetric slit compact CPMA.

6.2.5 SHORTING WALLS/PINS

The geometry of proposed virtually shorted patch antenna is shown in Figure 6.8. This antenna consists of four L-shaped parasitic shorting strips (L = 5 mm, t = 3.18 mm), a driven patch (W_p = 11.364 m), a ground plane (W_g = 21 mm), and four slots (S = 4.13 mm) and the patch is printed on the substrate. At all the corners of a square patch, four slots are loaded which are open-ended. These slots provide not only space for embedding the parasitic shorting strips but also size reduction for the patch. At each slot, strip is arranged individually. One end of each strip is open-circuited and other end of the strip is shorted to the ground plane by a shorting pin. The vertical shorting portion of each strip acts as an inductive loading whereas the horizontal portion of each strip is considered as a capacitive loading. The shorting strips are symmetrically arranged to maintain good broadside radiation. Circular polarization is obtained by controlling the extended length ratio of two orthogonal modes of the patch. To excite an circular polarization radiation, pairs of unbalanced tails are attached to the open ends of the slots as shown in Figure 6.8. At the center of the patch, a radiator designed with a metallic vertical post at DC ground is connected to the ground plane. The operating frequency is designed at 2.492 GHz. With the help of printed circuit board technique, antenna is fabricated. The fabricated antenna is realized by a PCB technique with the use of a dielectric substrate from Taconic with a thickness of ϵ =3.18 mm.

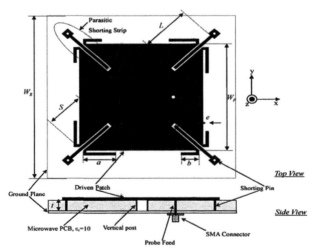

FIGURE 6.8 Antenna geometry for a circularly polarized patch antenna with parasitic shorting strips.

21. Wang, D.; Wong, H.; Chan, C. H. In *Small Circularly Polarized Patch Antenna*. International Workshop on Antenna Technology (iWAT), 2011; pp 271–273.

22. Ming Mak, K. A.; Lai, H. W.; Luk, K. M.; Chan, C. H. Circularly Polarized Patch Antenna for Future 5G Mobile Phones. *IEEE Antennas Wireless Propag. Lett.* **2015,** *2,* 1521–1529.

23. Garg, R.; Bhartia, P.; Bahl, I.; Ittipboon, A. *Microstrip Antenna Design Handbook*; Artech House: Norwood, MA, 2001.

24. Wong, H.; So, K. K.; Ng, K. B.; Luk, K. M.; Chan, C. H.; Xue, Q. Virtually Shorted Patch Antenna for Circular Polarization. *IEEE Antennas Wireless Propag. Lett.* **2010,** *9,* 1213–1216.

CHAPTER 7

HYBRID BEAM STEERABLE PHASED ARRAY ANTENNA FOR SATCOM OTM

V. DEVIKA[1,2*], K. SARAT KUMAR[2], K. CH. SRIKAVYA[1], AKHIL[3], and PRAGNYA[3]

[1]Department of ECE, KL University, Vijayawada, Andhra Pradesh, India

[2]Department of ECE, Hyderabad Institute of Technology and Management, Hyderabad, Telangana, India

[3]IV ECE, Department of ECE, Hyderabad Institute of Technology and Management, Hyderabad, Telangana, India

*Corresponding author. E-mail: Devikasv.ece@Gitam.org

ABSTRACT

Nowadays there is a great demand for communication on-the-move (OTM). OTM antenna is a system which is mounted on a vehicle such as boat, train, car, flight, and this system is used to track the satellite link and maintain the link between the terminal and satellite even when the vehicle is moving. The antenna always steers to track the satellite link while in motion. Phased array antenna with hybrid beam steering method is proposed in this paper to achieve effective communication. The beam formed in phased array of 100 isotropic elements has a widebeam width. But, to fetch satellite communication applications, a narrow beam width is required. Hence a parabolic reflector is chosen to sharpen the beam. But, for the ease of simulation, in place of the array of antennas, a horn antenna design to operate at the same operating frequency, 16 GHz, is given as a feed to the reflector antenna and the simulation is done using high-frequency structure simulator.

7.1 INTRODUCTION

Typically, satellite applications employ highly directive antennas, that is, parabolic dishes that are not suitable for mobile applications, where low-profile antennas are required.[7] To avoid complexity in antenna structures, we go for low-profile antennas. The phased array antennas are the best option for satellite communications (SATCOM) applications. To maintain the satellite link continuously on the move, the steering of the antenna plays a major role.

7.1.1 ACTIVE PHASED ARRAY ANTENNA

Phased array antennas are widely used in the typical satellite communication applications to achieve narrow beamwidth. Communication on the move prefers this phased array to achieve continuous link, coverage, and efficient data transmission. Phased array antenna is a multiple-antenna system in which the radiation pattern can be reinforced in a particular direction and suppressed in undesired directions. The direction of phased array radiation can be electronically steered obviating the need for any mechanical rotation. These unique capabilities have found phased arrays a broad range of applications since the advent of this technology. Phased arrays have been traditionally used in military applications for several decades. Recent growth in civilian radar-based sensors and communication systems has drawn increasing interest in utilizing phased array technology for commercial applications. Phased array antennas are common in communications and radar and offer the benefit of far-field beam shaping and steering for specific, agile operational conditions. They are especially useful in modern adaptive radar systems where there is a trend toward active phased arrays and more advanced space–time adaptive signal processing. In phased arrays, all the antenna elements are excited simultaneously and the main beam of the array is steered by applying a progressive phase shift across the array aperture.

Phased array antennas[1,8] have many features that would be beneficial for satellite communications on-the-move (SOTM). For example, the beam could be steered rapidly by electronically phase shifting the input signals (the so-called inertia-less beam), enabling the use of a high-speed scan without mechanical motion to estimate the pointing error. Alternatively, a multibeam phased array could also be configured to operate in a monopulse

7.2   HYBRID BEAM STEERING

Hybrid beam steering[3,5] is used in this phased array. For on-the-move (OTM) communications, it uses both electronically for elevation and mechanically for azimuth. This solution allows grouping the radiating elements by rows, defining a set of subarrays, each one controlled by a phase shifter device. The active phase array antenna will be mounted on a stabilized platform that allows dynamic compensation for roll, pitch, and yaw in "on-the-move" applications.

The antenna tracking system controls the beam pointing, electronically in elevation and mechanically in azimuth, to keep the satellite link active. Due to considerable distance between the transmitter and receiver Ka-bands, two different phased array systems have to be developed. However, the higher frequencies involved in Ka-band allow reducing antenna dimensions and each phased array is optimized in terms of antenna and beam forming network design. The single radiating element is constituted by a couple of slot on a waveguide, opportunely oriented to generate a circular polarized field.

The antenna system is completed with a stabilized platform and an antenna tracking system (ATS). The stabilized platform is able to provide the required pointing angles stabilization with respect to the moving ground vehicle. The platform is made up by a fixed and a rotating part. The fixed part includes all the sensors needed to measure the attitude and the position of the vehicle. The ATS collects the position and attitude information and then drives the mechanical and electrical axes to provide the required stabilization.

7.3   METHODOLOGY

There are two broad approaches for this system.

1. Open-loop approach
2. Closed-loop approach

In open-loop approach,[2] the antenna is oriented toward the known position of the geostationary satellite. This approach depends on the inertial movement of the vehicle on which antenna is placed. As vehicle is on the move, the antenna will reorient to current position to maintain the link with the satellite. During this process, there is a chance of occurrence of pointing error. Pointing accuracy cannot be achieved within a fraction of degrees in this open loop system due to its dependence on inertial measurement system to steer the antenna.

In the closed-loop approach,[2] the antenna tracks the satellite link by considering the strongest receiver signal or beacon signal from the satellite's own transmission. To find the maximum signal strength, mechanical scanning of conventional reflector antenna across the sky is required. A deliberate pointing error is to be introduced to verify or check the maximum receiver signal strength. Due to this, the system responds too slowly for speedy vehicle movements. Instead, monopulse tracking system[4] can be used to find the accuracy in finding the precise direction. This system has ability to estimate pointing error without any mechanical scanning and without deliberately mispointing. Dual feed method is used in this monopulse antennas; one feed generates normal radiation pattern of the antenna, while the other feed internally generates radiation pattern with a sharp notch along the bore sight. The output signals from two feeds are compared and antenna can be precisely pointed to eliminate pointing error.

The pointing error[2] is generally less than 0.1° over a full elevation coverage, which indicates the perfect communication link even in demanding motion environment (Fig. 7.2).

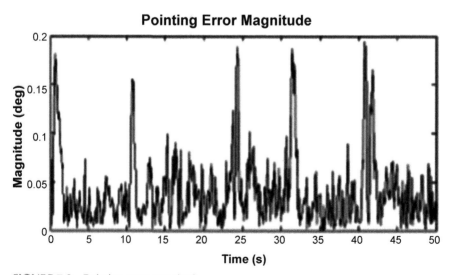

Pointing Error Magnitude

FIGURE 7.2 Pointing-error magnitude.

7. Tripodi, M.; et al. In *Ka Band Active Phased Array Antenna System for Satellite Communication on the Move Terminal*. Proceedings of the 5th European Conference on Antennas and Propagation (EUCAP); 2011.

8. Mailloux, R. J.; Books, I. *Phased Array Antenna Handbook*; Artech House: Norwood, MA, 2005.

CHAPTER 8

SUPERSTRATE-LOADED SQUARE-PATCH ANTENNA ANALYSIS

V. SAIDULU*

Department of ECE, MGIT, Hyderabad 500075, Telangana, India

Corresponding author. E-mail: saiduluvadtya@gmail.com

ABSTRACT

This paper describes the effect of the superstrates on the performance characteristics of square patch microstrip antenna without and loaded with dielectric superstrates. It is found that there is a degradation in the performance of the antenna when the superstrate is touching the patch antenna, that is, its height above the patch antenna $(H) = 0$ mm. Further, it is also observed that the degraded performance characteristics of the patch antenna can be improved by placing the superstrates at optimum height $(H) = H_{opt}$. The microstrip patch antenna without dielectric superstrate has an impedance bandwidth of 0.041GHz (SWR ≤ 2) at 2.40 GHz, gain is 8.90 dB, and return loss is -23.00 dB. When the superstrate is placed touching the patch antenna, the resonant frequency is reduced to 2.35 GHz from 2.40 GHz and bandwidth is reduced 0.032 GHz (SWR ≤ 2) at 2.35 GHz, gain is decreased by about 3.37% (0.3 dB) for $\epsilon_r = 2.2$ to 28% (2.5 dB) for $\epsilon_r = 10.2$. As the height of the superstrate is increased, the performance of the patch antenna improves, and at a particular optimum height, the gain and band width for all the superstrates will be closer to the free space radiation conditions of the patch antenna without superstate. But the resonant frequency decreases with increase in ϵ_{r2}. There is a good agreement between simulated and measured results.

8.1 INTRODUCTION

Square-patch antenna employed for various applications like aircraft, spacecraft, satellite, and missile, where size, weight, cost, and aerodynamic profile are constraints. This chapter presents the effect of the superstrate on the characteristics of square-patch antenna. The schematic diagram of the patch antenna loaded with superstrate is shown in Figure 8.1.

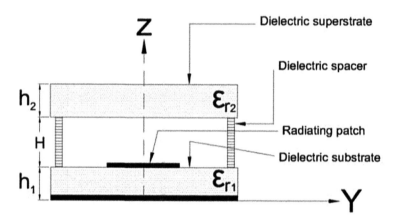

FIGURE 8.1 The schematic of a patch antenna loaded with a superstrate at height (H) above the patch (side view).

The dielectric superstrates of different dielectric constants are used to study the effect on the performance of the patch antenna. The height of the superstrate is varied and the effect of the height is investigated. The simulation method using high-frequency structure simulator (HFSS), version 13.0, is employed to obtain the simulated results of performance characteristics without superstrate and loaded with superstrates as a function of dielectric constant and height of the superstrate. HFSS is used due to their simplicity and they make the design easy.[1–15] The experimental results are obtained with the help of Precision Network Analyzer (Agilent E8363B) and anechoic chamber.

8.2 SPECIFICATIONS

The dielectric constants, loss tangents, and thicknesses of the dielectric materials used in the investigations are given in Tables 8.1 and 8.2. Dielectric substrate of appropriate thickness and loss tangent is chosen for designing the square microstrip patch antennas (MPAs). A thicker substrate is mechanically strong with improved impedance bandwidth and gain. However, it also increases weight and surface wave losses. The dielectric constant ($\in_r$) plays an important role similar to that of the thickness of the substrate. A low value of $\in_r$ for the substrate will increase the fringing field of the patch and thus the radiated power. A high loss tangent (tan δ) increases the dielectric loss and therefore reduces the antenna performance. The low dielectric constant materials increase efficiency, bandwidth, and radiation.[2–3,5–6,8,12–14]

Keeping these aspects in mind, the square MPAs are fabricated on Arlon DiClad 880 dielectric substrate, whose dielectric constant ($\in_{r1}$) is 2.2, loss tangent (tan δ) is 0.0009, thickness (h_1) is 1.6 mm, and which has appropriate substrate dimensions.

TABLE 8.1 Specification of Dielectric Substrate ($\in_{r1}$) Material Used in the Design of Patch Antenna.

Substrate material	Dielectric constant ($\in_{r1}$)	Loss tangent (tan δ)	Thickness of the substrate (h_1)(mm)
Arlon DiClad 880	2.2	0.0009	1.6

TABLE 8.2 Specification of Dielectric Superstrate ($\in_{r2}$) Materials Used to Study the Effect of the Superstrate on the Performance of the Antenna.

Superstrate materials	Dielectric constant ($\in_{r2}$)	Loss tangent (tan δ)	Thickness of the superstrates (h_2) (mm)
Arlon DiClad 880	2.2	0.0009	1.6
Arlon AD 320	3.2	0.003	3.2
FR4	4.8	0.02	1.6
Arlon AD 1000	10.2	0.0035	0.8

8.3 DESIGN OF SQUARE-PATCH ANTENNA AND THEIR GEOMETRY

Square microstrip patch antenna is formulated using the transmission line model and designed at the center frequency of 2.4 GHz on Arlon DiClad 880 substrate ($\in_{r1} = 2.2$, $h_1 = 1.6$ mm). The designed dimensions of square-patch antennas are given in Table 8.3. The patch antennas are fed with coaxial probe feed at a point where the input impedance of the patch is 50 Ω.

TABLE 8.3   The Measured Dimensions of the Square-patch Antenna (in mm).

W$_p$	L$_p$	F$_X$,F$_Y$
40.30	40.30	10.0

The location coordinates (F_X, F_Y) are found by simulation. The geometries of the square-patch antennas are shown in Figure 8.2. In the geometry shown, W_p is the patch width, L_p is the patch length, and (F_X, F_Y) are the coordinates of the feed point.

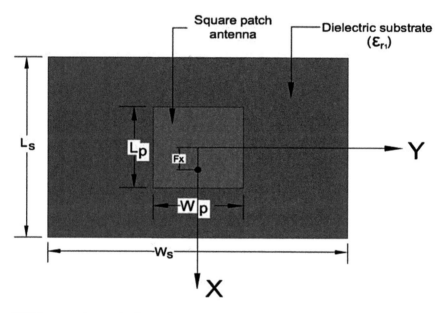

FIGURE 8.2   (See color insert.) Geometry of square-patch antenna (top view).

8.4   SIMULATED AND EXPERIMENTAL RESULTS

The performance characteristics of the square-patch antenna are evaluated without dielectric superstrate using commercial electromagnetic software such as HFSS, version 13.0. Then, the change in performance of the antenna is studied with dielectric superstrate of dielectric materials as mentioned in Table 8.2. The effect of the height of the superstrate above the patch (H) is also studied by simulation. The height at which the performance of the patch is optimum is also found by simulation using HFSS version 13.0.

The measurements were carried out by using Precision Network Analyzer (Agilent E8364B) to measure the return loss voltage standing wave ratio (VSWR), center frequency, and bandwidth and Anechoic chamber to measure the radiation characteristics. The antenna under test (patch antenna with and without dielectric superstrate) is used as receiving antenna and the transmitting antenna is a pyramidal horn antenna (0.5–6 GHz). The antenna measurements were carried out in Anechoic chamber having dimensions (30 × 20 × 15 ft). The distance between transmitting and receiving antenna is kept as 5.3 m. The radiation pattern measurements were carried out at 2.4 GHz.

8.5 RESULTS AND DISCUSSION

8.5.1 RESULT OF SQUARE-PATCH ANTENNA WITHOUT SUPERSTRATE

The simulated and measured results of return loss and radiation patterns in E-plane and H-plane for the square-patch antenna under free-space radiation conditions, that is, without superstrate, are shown in Figures 8.3 and 8.4.

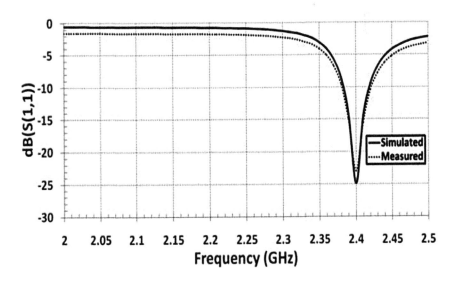

FIGURE 8.3   Comparison of measured and simulated results of return loss for square microstrip patch antenna without dielectric superstrate $\epsilon_{rl} = 2.2$ (free-space radiation conditions).

From Figure 8.4, it can be observed that there is a good agreement between simulated and measured results.[5] The resonant frequency is 2.40 GHz, same as the design frequency, the bandwidth is 0.041 GHz (VSWR ≤ 2), and the gain is 8.90 dB in both cases.

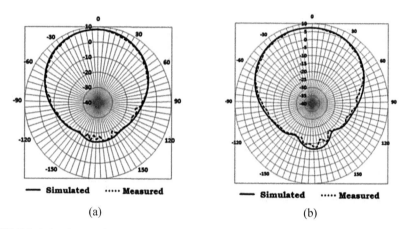

(a)   (b)

FIGURE 8.4 Comparison of measured and simulated results of radiation patterns for square microstrip patch antenna in (a) E-plane and (b) H-plane for $\epsilon_{rl} = 2.2$ without dielectric superstrate (free-space radiation conditions) at 2.40 GHz.

Radiation pattern simulation and measurements are carried out at the center frequency for the case under consideration.[5] The center frequency is found from return-loss measurements.

For free-space radiation conditions, that is, without superstrate, the center frequency is occurring at 2.40 GHz, and hence, the radiation pattern simulation and measurements are carried out at this frequency, that is, at 2.40 GHz.

Figure 8.5 shows the axial ratio versus frequency plot. The axial ratio is >50 dB (AR > 100) over the operating frequencies (2.35–2.45 GHz). This indicates that the square-patch antenna produces linear polarization. There is a good agreement between simulated and measured results.

8.5.2 RESULT OF SQUARE-PATCH ANTENNA WITH DIELECTRIC SUPERSTRATES

The simulation for various dielectric superstrates as a function of height (*H*) on the performance characteristics of square MPA has been carried

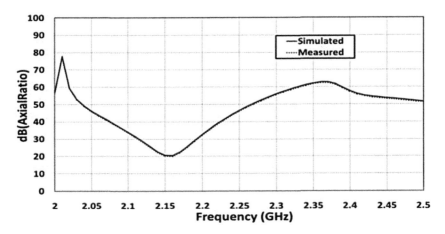

FIGURE 8.5   Comparison of measured and simulated results of axial ratio versus frequency plot for square-patch antenna without dielectric superstrate $\epsilon_{r2} = 2.2$ (free-space radiation conditions).

out as mentioned in Table 8.4 shows overall comparison of simulated and measured results of resonant frequency, return loss, bandwidth, gain, and VSWR of square microstrip patch antenna without superstrate and loaded with superstrates. The simulated results indicate that the behavior of a square MPA is similar to that of the rectangular MPA, when loaded with a dielectric superstrate. The optimum height is found to be same in both cases for superstrate of a particular dielectric constant. Measurements have been carried out for various dielectric superstrate $\epsilon_{r2} = 2.2, 3.2, 4.8, 10.2$. Typical case is $\epsilon_{r2} = 2.2$; results are discussed below.

Superstrate with $\epsilon_{r2} = 2.2$ and $h_2 = 1.6$ mm.

The effect of the superstrate having $\epsilon_{r2} = 2.2$, $h_2 = 1.6$ mm on the performance characteristics of the square patch is evaluated using simulation and measurements.

The simulation and measurements are carried out for $H = 0$, and $H = H_{opt}$. The simulated and measured results are shown in Figures 8.6–8.9. The plot of return loss as a function of frequency is shown in Figures 8.6 and 8.7. The plot of radiation patterns is shown in Figures 8.8 and 8.9. The radiation patterns are measured at the resonant frequency for the case under consideration. It is found that there is a good agreement between simulated and measured results.

TABLE 8.4 The Overall Comparison of Simulated and Measured Results of Resonant Frequency, Return Loss, Bandwidth, Gain, and VSWR of Square Microstrip Patch Antenna Without Superstrate and Loaded with Superstrates.

H_{opt}	Height (H) (mm)	Frequency (GHz)		Return loss (dB)		Bandwidth (GHz)		Gain (dB)		VSWR	
		Simulated	Measured	Simulated	Measured	Simulated	Measured	Simulated	Measured	Simulated	Measured
1	–	2.40	2.40	−24.94	−23.00	0.040	0.041	8.90	8.90	1.12	1.15
2.2	0	2.35	2.35	−24.30	−23.20	0.031	0.032	8.74	8.60	1.13	1.15
	21.07 (H_{opt})	2.40	2.40	−19.61	−17.30	0.041	0.042	8.80	8.70	1.23	1.32
3.2	0	2.32	2.32	−26.77	−25.00	0.031	0.033	8.32	8.30	1.10	1.12
	17.46 (H_{opt})	2.41	2.41	−17.30	−16.90	0.040	0.041	8.66	8.60	1.32	1.33
4.8	0	2.27	2.27	−25.12	−23.60	0.032	0.032	7.67	7.60	1.12	1.14
	14.26 (H_{opt})	2.41	2.41	−15.64	−16.40	0.041	0.042	8.71	8.70	1.40	1.36
10.2	0	2.12	2.12	−21.34	−20.50	0.022	0.023	6.44	6.40	1.19	1.21
	9.78 (H_{opt})	2.41	2.41	−12.24	−11.70	0.040	0.041	8.80	8.80	1.65	1.70

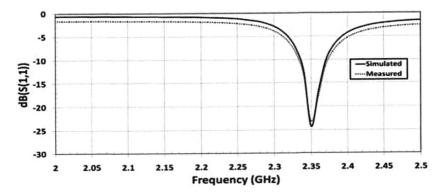

FIGURE 8.6   Comparison of measured and simulated results of return loss for square-patch antenna loaded with a dielectric superstrate $\epsilon_{r2} = 2.2$, $h_2 = 1.6$ mm) for $H = 0$.

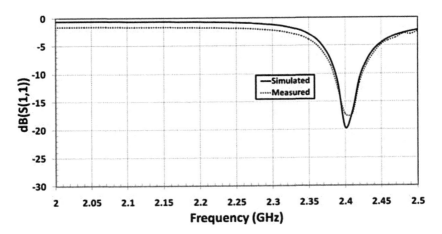

FIGURE 8.7   Comparison of measured and simulated results of return loss for square-patch antenna loaded with a dielectric superstrate $\epsilon_{r2} = 2.2$, $h_2 = 1.6$ mm) for $H = H_{opt}$ at 21.07 mm.

For $H = 0$, the center frequency is occurring at 2.35 GHz and hence the radiation pattern simulation and measurements are carried out at 2.35 GHz. For $H = H_{opt}$, the center frequency is occurring at 2.40 GHz and hence the radiation pattern simulation and measurements are carried out at 2.40 GHz.

When the superstrate is touching the patch antenna ($H = 0$), the measured values of f_r, bandwidth, and gain are observed to deteriorate as found in the case of simulation. The resonant frequency is decreased to 2.35 GHz, the bandwidth is decreased to 0.032 GHz, and gain is decreased to 8.60 dB.

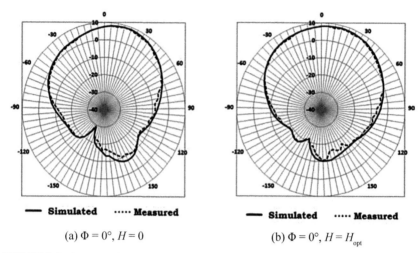

(a) $\Phi = 0°$, $H = 0$ (b) $\Phi = 0°$, $H = H_{opt}$

FIGURE 8.8 Measured and simulated radiation patterns of square-patch antenna in E-plane for $\in_{r2} = 2.2$, $\Phi = 0°$: (a) $H = 0$, $f_r = 2.35$ GHz and (b) $H = H_{opt}$, $f_r = 2.40$ GHz.

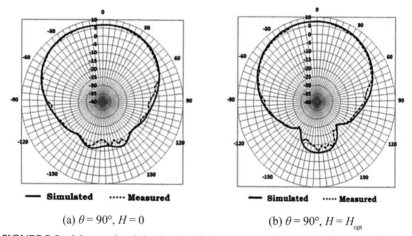

(a) $\theta = 90°$, $H = 0$ (b) $\theta = 90°$, $H = H_{opt}$

FIGURE 8.9 Measured and simulated radiation patterns of square-patch antenna in H-plane for $\in_{r2} = 2.2$, $\theta = 90°$: (a) $H = 0$, f_r 2.35 GHz and (b) $H = H_{opt}$, f_r 2.40 GHz.

When the superstrate is kept at $H = H_{opt} = 21.07$ mm, the f_r, bandwidth, and gain are found to improve. Both simulation and measurements have been carried out for $H = H_{opt}$ and they are found to be in good agreement. The measured values of resonant frequency are 2.40 GHz, the bandwidth is 0.042 GHz, and gain is 8.70 dB.

As compared to free-space radiation conditions, for $H = 0$, the resonant frequency (f_r) is 2.35 GHz (2.08% less), bandwidth is 0.032 GHz (21.95% less), and gain is 8.60 dB (3.37% less). For $H = H_{opt} = 21.07$ mm, the resonant frequency (f_r) is 2.40 GHz (same as without superstrate), bandwidth is 0.042 GHz (2.43% more), and gain is 8.70 dB (2.24% less). Hence, it can be concluded that when $H = H_{opt} = 21.07$ mm, the performance characteristics will be closer to the values measured under free-space radiation conditions. The effect of the superstrate ($\in_{r2}$ = 2.2) is found to be negligible on the axial ratio for all heights. The overall comparison of simulated and measured results of resonant frequency, return loss, bandwidth, gain, and VSWR of square microstrip patch antenna without superstrate and loaded with superstrates is shown in Table 8.4.

8.6 CONCLUSIONS

Square microstrip patch antenna has been designed and fabricated at 2.4 GHz with Airlon DiClad 880 substrate having $\in_{r1}$ = 2.2. The effect of the superstrate with different dielectric materials having $\in_{r2}$ = 2.2, 3.2, 4.8, and 10.2 has been investigated. The simulation and measurements have been carried out for studying the effect of superstrates on various parameters like resonant frequency, bandwidth, gain, and return loss. It has been observed that there is a degradation in the performance of the antenna when the superstrate is touching the patch antenna ($H = 0$). The center frequency is decreased to 2.35 GHz from 2.4 GHz (2.08%), bandwidth is decreased to 0.032 GHz from 0.041 GHz (21.90%), and gain is decreased to 8.60 dB from 8.90 dB (3.37%) for $\in_{r2}$ = 2.2. As height of the superstrate is increased the performance of patch antenna improves at the optimum height (H_{opt}). The center frequency is 2.40 GHz (same as without superstrate), bandwidth is 0.042 GHz, and gain is 8.70 dB for $\in_{r2}$ = 2.2, which is closer to free-space radiation conditions of the patch antenna without superstrate. Similarly for other dielectric constants of the superstrate the result is shown in Table 8.4. But the resonant frequency decreases as dielectric constant of the superstrate ($\in_{r2}$) increases. The simulated and measured results are in good agreement.

ACKNOWLEDGMENTS

Author acknowledges the invaluable help of Mr. M. Balachary, Scientist 'G' and Head of Antenna Wing, DLRL, Hyderabad who helped greatly by offering the experimental measurement work at DLRL.

KEYWORDS

- **square-patch antenna**
- **superstrates**
- **dielectric constant**
- **transmission line model**
- **patch antenna**
- **high-frequency structure simulator**
- **precision network analyzer**

REFERENCES

1. Munson, R. E. Conformal Microstrip Phased Arrays. *IEEE Trans. Antennas Propag.* **1974,** *AP-22,* 74–78.
2. Bhal, I. J.; Bhartia, P. *Microstrip Antenna*; Artech House: Boston, MA, 1980.
3. Balanis, C. A. *Antenna Theory: Analysis and Design*, John Wiley & Sons: Hoboken, NJ, 2016.
4. Agrawal, P. K.; Bailey, M. C. An Analysis Technique for Feed Line Microstrip Antennas. *IEEE Trans. Antennas Propag.* **1977,** *AP-25,* 756–758.
5. Yadav, R. K.; Yadava, R. L. Superstrate Loaded Rectangular Microstrip Antennas—An Overview. *JIIK* **2011,** *3* (2), 19–36.
6. Bahl, J.; Stuchly, S. S. Analysis of Microstrip Covered with a Lossy Dielectric. *IEEE Trans.* **1980,** *MTT-28,* 104–109.
7. Pues, H.; Van de Capelle, A. Accurate Transmission Line Model for the Rectangular Microstrip Antenna. *IEEE Proc.* **1984,** *131,* 334–340.
8. Shavit, R. Dielectric Cover Effect on Rectangular Microstrip Antenna Array. *IEEE Trans. Antenna Propag.* **1994,** *AP-42,* 1180–1184.
9. Meagher, C. J.; Sharma, S. K. A Wide Band Aperture Coupled Microstrip Patch Antennas Employing Space and Dielectric Cover for Enhanced Gain Performance. *IEEE Trans. Antenna Propag.* **2010,** *58* (9), 2802–2810.
10. Attia, H.; Yousefi, L.; Ramahi, O. M. Analytical Model for Calculating the Radiation Fields of MSA with Artificial Magnetic Superstrates: Theory and Experiment. *IEEE Trans. Antennas Wave Propag.* **2011,** *59,* 1438–1445.
11. Tamboli, Z. J.; Nikam, P. B. A Study of Multilayer Perceptron Neural Network for Antenna Characteristics Analysis. *Int. J. Adv. Res. Comput. Sci. Softw. Eng.* **2013,** *3* (8), 200–203.
12. Saidulu, V.; Srinivasa Rao, K. Experimental Studies on Microstrip Patch Antenna with Superstrate. In *Proceeding of National Conference on Recent Advancement in Electronics (NCRAE)*, Faculty of Science and Technology, IFHE University, Hyderabad, January 2016.

13. Saidulu, V.; Srinivasa Rao, K. Study of the Dielectric Superstrate Thickness Effects on Microstrip Patch Antenna. *IOSR J. Electron. Commun. Eng.* **2016,** *11* (1), 55–65.
14. Saidulu, V.; Srinivasa Rao, K. Analogy of Microstrip Patch Antenna with Superstrate. In *Proceeding of International Conference on Innovations and Advancements in Computing (ICIAC-2016)*, GITAM University, Hyderabad, March 2016; pp 207–213.
15. Bahl, I. J.; Bhatiya, P.; Stuchly, S. S. Design of Microstrip Antenna Covered with a Dielectric Layer. *IEEE Trans.* **1982,** *AP-30*, 314–318.

CHAPTER 9

IMPLEMENTATION OF GFDM TRANSCEIVER

K. PRUTHVI KRISHNA, SHRAVAN KUMAR BANDARI, and V. V. MANI*

Department of Electronics and Communication Engineering, National Institute of Technology, Warangal, Telangana, India

Corresponding author. E-mail: vvmani@nitw.ac.in

ABSTRACT

Generalized frequency division multiplexing (GFDM) is a nonorthogonal multicarrier multiplexing scheme. GFDM is new physical level modulation scheme aimed at replacing the orthogonal frequency division multiplexing modulation scheme, thus providing better and alternative approach, to be implemented in upcoming higher versions of wireless communication systems such as 5G. The software used for implementing GFDM is LabVIEW. LabVIEW is system design platform environment that supports graphical programming and also has easy interface with real-time devices. In this paper, the main aim is to design a complete transceiver model, and check both transmitter receiver constellations. The proposed system design performance is checked by transmitting and receiving text message under different additive white Gaussian noise environment.

9.1 INTRODUCTION

5G refers to one of the major phase changes in the mobile telecommunication standards which goes beyond 4G or long-term evolution (LTE) advanced. It aims at providing better data rates, more coverage, and enhanced signaling efficiency at much improved latency. However, the main problem is that present generation techniques and methods won't be sufficient enough to achieve the desired characteristics. The scenarios foreseen for future fifth-generation (5G) networks have requirements that clearly go beyond higher data rates which are being used at present in 3G or 4G. The main scenarios for 5G networks are machine-type communication, tactile Internet, and wireless regional area network, while classical bit pipe communication is still considered an important application.

Generalized frequency division multiplexing (GFDM) is basically a physical (PHY) layer concept taken as a replacement for the existing PHY layer technique called as orthogonal frequency division multiplexing (OFDM). OFDM cannot be used in future generation system due to its synchronization requirements to maintain the orthogonality at the transceiver and high peak-to-average power ratio requirements. Furthermore, it is not suitable for higher generation systems because of its spectral leakage and out of band emissions. Also, using OFDM, it is not possible to attain the high data rates and bandwidth that are necessary for 5G.

The major requirements for 5G are as follows:

- Data rates up to 1 GB/s.
- Better spectrum utilization.
- Reduced latency in comparison to LTE.
- Better synchronization with IoT devices.

The following contains a brief review of literature regarding implementation of GFDM. 5G demands for higher data rates that exceed the present generation capabilities of low power consumption and many other factors; thus, they proposed new PHY layer technique, referred to as GFDM, to meet the above requirements and thereby giving the principles of GFDM.[1]

Cognitive radio system requires techniques that have low out of band emission. They felt that OFDM is not that suitable as it doesn't reduce the out of band emission to the extent required. However, they found a technique called GFDM which can be used as a replacement for OFDM. Thus, they studied this technique performance and compared with existing OFDM.[2]

GFDM can be seen as generalization of traditional OFDM. Thus, this scheme can be implemented with less computational effort using fast Fourier/inverse Fourier transform (FFT/IFFT) algorithm.[3]

Due to the presence of nonorthogonal carriers in GFDM, the signal can be distorted and effected because of intersymbol interference and intercarrier interference. Thus, they discovered that a small addition of orthogonality could increase the overall performance of the GFDM.[4]

GFDM is a generalized digital multicarrier transceiver concept. According to them, GFDM is a digitally implemented traditional filter bank multibranch multicarrier concept which doesn't need synchronization due to nonorthogonality of carriers and low out of band emissions, thus making suitable for higher generation cellular communication systems.[5] Meyer function can be represented as root raised cosine (RRC) filter when the fractional excess bandwidth is taken above by one-third. Thus, the Meyer function can be used as RRC filter in the software simulation process.[6]

The goal of the chapter is to implement GFDM successfully on system design platform environment known as LabVIEW. LabVIEW is a graphical programming language which can be easily interfaced with real-time devices and also allow programming using mathscript and other languages like C. Thus in short, the aim of this chapter is to implement GFDM in LabVIEW and check its performance characteristics. The rest of the chapter is organized as follows: The general block diagram and mathematical representation of GFDM transmitter and receiver is presented in Section 9.2. Results of the proposed system and its performance are discussed in Section 9.3, followed by conclusion and future work in Section 9.4.

9.2   SYSTEM MODEL

The generic block diagram of GFDM is as shown in Figure 9.1.

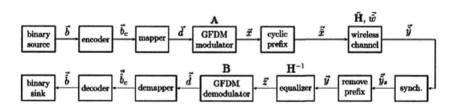

FIGURE 9.1   Block diagram of GFDM transceiver as in Ref. [1].

9.2.1   TRANSMITTER

GFDM is a multicarrier modulation scheme that has been introduced by Fettweis.[5] Figure 9.2 depicts the block diagram of the GFDM transmitter. The input random bits in the form of data streams are fed to K independent mappers. Each mapper converts a block of $\log_2(J)$ bits into a data symbol that can be transmitted in K separate subcarriers. Different order and even different modulation schemes can be used for different streams as the mappers used are mutually exclusive from one another. In a GFDM, M data symbols are transmitted within the same subcarrier using M time slots.

The modulator function is explained in detail using Figure 9.2.

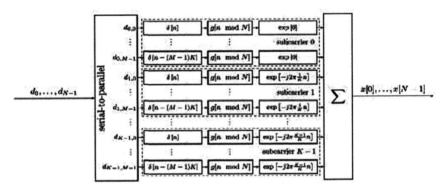

FIGURE 9.2   GFDM modulator as in Ref. [1].

As it can be seen through Figure 9.2, the data symbols are distributed across K active subcarriers. These data symbols within each subcarrier are further distributed across M active subsymbols and each subcarrier is pulse shaped using a RRC filter whose filter coefficients are denoted by $\tilde{g}_{Tx}[n]$ and are then modulated with a subcarrier center frequency $e_{j2\_n=N}$. Each symbol is sampled N times which should be greater than or equal to the total number of subcarriers, that is, total of MN samples per subcarrier, which is required to satisfy the Nyquist criterion.

Thus, the transmitted signal will be given by as in Ref. [2].

$$x[n] = \sum_{m=0}^{M-1} \sum_{k=0}^{K=0} d_k(m) g_{Tx}[n-mN] e^{j2\pi n/N} \tag{9.1}$$

Note that the filter $g_{Tx}[n]$ is considered as circular with a period of n mod mN so as to reduce the guard time interval, thereby contributing it to better spectrum utilization. This method is known as tail biting mechanism in which with the use of circular convolution, the last mN samples are shifted to first mN positions.

In vector form, it can be represented as

$$\mathbf{x} = \mathbf{Ad} \tag{9.2}$$

where $\mathbf{A}$ denotes an NM–KM modulation matrix. The matrix contains the responses of the pulse shaping filter for all possible time and frequency shifts.

However, it can be seen that the computations required using a matrix or vector form are very high, thus making it very complex to implement. But one of the easiest ways to implement GFDM in real time without hefty calculations is use of FFT/IFFT algorithm similarly to that used in OFDM. Thus, the transmitted signal in Equation (9.1) can be represented as in Ref. [1].

$$x_k[n] = \left[\left(d_k[m]\delta[n-mN] \right) \times g_{Tx}[n] \right] e^{j2\pi n/N} \tag{9.3}$$

where x_k denotes the transmit signal of the kth subcarrier. The modulation of an individual subcarrier in (9.3) can be broken down to the convolution of a Dirac pulse train ($d_k[m]\delta[n - mN]$) with a filter response $g_{Tx}[n]$ and a subsequent multiplication with a complex valued oscillation $e^{j2\pi n/N}$.

It can be easily solved when done in frequency domain. However, since each subcarrier carries M subsymbols, there exists a possibility of symbols overlapping with each other which leads to intersymbol interference. Also, the pulse-shaping filter used is not rectangular; thereby, the subcarriers are not orthogonal anymore as they are in the case of OFDM, thus causing intercarrier interference. Nonorthogonality of subcarrier results in both merits and demerits of its own as nonorthogonality results in less synchronization but makes it vulnerable at receiver side due to incorrect reception of data. Thus to overcome this demerit, cyclic prefix (CP) is used. CP concept is similar to that of OFDM in which the last X input bits are copied and placed in the header to the front. However, header reduces interframe interference and it is not required between every time slot; rather, it is used only in between frames, thus reducing the header overload. Thus, the GFDM signal is ready for transmission.

However, assume the channel to be additive white Gaussian noise (AWGN) and add noise coefficients to it with a constant E_b/N_0.

The GFDM signal can be received and the original signal can be reconstructed by various methods like zero-forcing receiver or matched filter, etc. One of the best methods using will be the use of matched filter. The system parameters considered for simulation are shown in Table 9.1.

TABLE 9.1 Specifications of the GFDM System Implemented.

Parameter	Value
Number of subsymbols per subcarrier	16
Number of subcarriers	32
Number of samples	64
Modulation scheme used	16-QAM
Cyclic prefix length	16 bits

9.2.2 RECEIVER

As specified in Section 9.2.1, the signal can be reconstructed using one of many methods. Here, matched filter is used as receiver in which K parallel receivers are used. Each of these receivers can be considered as a correlator receiver as shown in Figure 9.3. The received GFDM signal first goes through a CP remover. After removing CP, the whole process is reverse engineered using matched filter concept in which the received signal in frequency domain is multiplied with matched coefficient of root raise cosine filter per subcarrier in same circular convoluted manner as that in the transmitter to get J-QAM signals. These QAM symbols are then soft mapped which are further remapped into corresponding random bits. The constellation and remaining parameters of remapped bits are compared with the transmitter constellation with the corresponding parameters (Figs. 9.4–9.6).

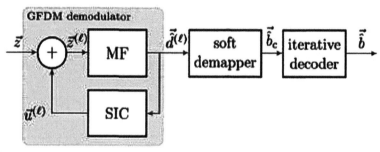

FIGURE 9.3   GFDM demodulator as in Ref. [1].

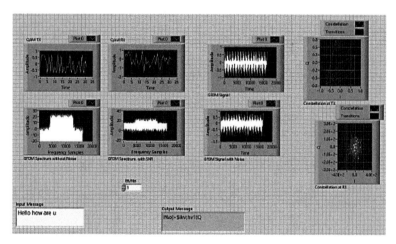

FIGURE 9.4   GFDM performance analysis for $E_b/N_0 = 1$ dB.

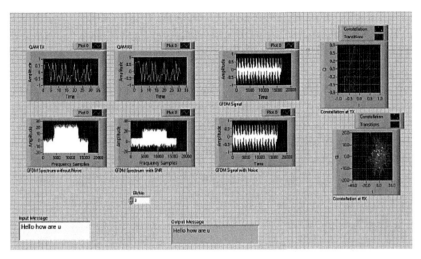

FIGURE 9.5   GFDM performance analysis for $E_b/N_0 = 3$ dB.

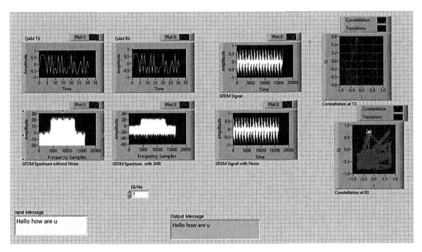

FIGURE 9.6   GFDM performance analysis for $E_b/N_0 = 7$ dB.

9.3   RESULTS

The performance of the implemented GFDM is observed by taking the different E_b/N_0 values and checking the corresponding constellations, GFDM signal before and after affected by AWGN noise, and their respective spectrum. The performance analysis is actually done by transmitting a constant text message using GFDM for different E_b/N_0 values and comparing it with the received text message. It has been observed that for the value of E_b/N_0 equal to 1 dB, the signal is totally unrecoverable but when E_b/N_0 is equal to 3 dB, the constellation of the symbol is not recoverable but sometimes text message is recovered fully and sometimes partially as it depends upon the soft remapping of QAM symbols. Thus, it can be said that up to 3 dB AWGN Gaussian noise, the signal cannot be recovered back. But as it can be seen when E_b/N_0 reaches the values of 7 dB, the signal is reached; the signal is recoverable along with the text message and also the received symbols and the corresponding spectrum, whereas in case of 9 dB E_b/N_0 value which is rather a very high value, the whole signal along with perfect constellation and text message is attained back (Fig. 9.7).

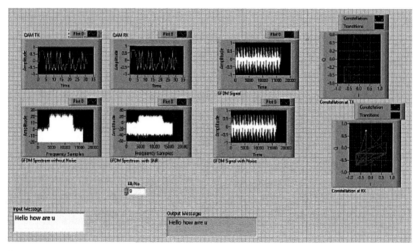

FIGURE 9.7   GFDM performance analysis for $E_b/N_0 = 9$ dB.

9.4   CONCLUSION AND FUTURE WORK

In this chapter, GFDM transceiver is implemented using LabVIEW in which the AWGN noise effect is taken on the account; on the basis of different values of $E_b = N_0$, the text message received is compared with the text transmitted. It has been noticed that at the lower values of $E_b = N_0$, that is, in the range less than 3 dB, the effect of noise is dominant on the signal and signal cannot be recovered back under any circumstances. But as the $E_b = N_0$ value increases gradually, it has been noticed that the signal can be received partially with slight mismatch of text in the range of 3–7 dB, given that the equalization is not taken into account. However, the signal can be recovered without any error for the $E_b = N_0$ values greater than or equal to 7 dB even without equalization. In the future, this work will be extended to real-time devices using USRP RIO SDR device and also channel estimation will be carried out along with the equalization at the receiver to give a working prototype of GFDM at some particular frequency.

KEYWORDS

- **mobile telecommunication standards**
- **pipe communication**
- **nonorthogonal carriers**
- **cellular communication systems**
- **long-term evolution**
- **high peak-to-average power ratio**
- **orthogonal frequency division multiplexing**

REFERENCES

1. Michailow, N.; Matthe, M.; Gaspar, I. S.; Caldevilla, A. N.; Mendes, L. L.; Festag, A.; Fettweis, G. Generalized Frequency Division Multiplexing for 5th Generation Cellular Networks. *Commun., IEEE Trans.* **2014,** *62* (9), 3045–3061.
2. Alves, B. M.; Mendes, L. L.; Guimaraes, D. A.; Gaspa, I. S. Performance of GFDM over Frequency-Selective Channels. In *International Workshop on Telecommunications Conference*, May 2013.
3. Michailow, N.; Gaspar, I.; Krone, S.; Lentmaier, M.; Fettweis, G. Generalized Frequency Division Multiplexing: Analysis of an Alternative Multi-carrier Technique for Next Generation Cellular Systems. In *Wireless Communication Systems (ISWCS), 2012 International Symposium on,* 28–31 Aug. 2012; pp 171–175.
4. Gaspar, I.; Matthe, M.; Michailow, N.; Leonel Mendes, L.; Zhang, D.; Fettweis, G. Frequency-Shift Offset-QAM for GFDM. *Commun. Lett. IEEE* **2015,** *19* (8), 1454–1457.
5. Fettweis, G.; Krondorf, M.; Bittner, S. GFDM – Generalized Frequency Division Multiplexing. In *Vehicular Technology Conference, 2009. VTC Spring 2009, IEEE 69th,* 26–29 April 2009; pp 1–4.
6. Jones, W. W.; Dill, J. C. The Square Root Raised Cosine Wavelet and Its Relation to the Meyer Functions. *Signal Process., IEEE Trans.* **2001,** *49* (1), 248–251.

PART II

Communication Systems

CHAPTER 10

ACHIEVABLE SUM SPECTRAL EFFICIENCY ANALYSIS OF MASSIVE MIMO WITH A MMSE-SIC RECEIVER

KRISHNA PATTETI[1*], M. SAMPATH REDDY[1],
ANIL KUMAR TIPPARTI[2], and K. SRINIVASA RAO[3]

[1]*Department of Electronics and Communication Engineering, Jayamukhi Institute of Technological Sciences, Warangal, Telangana, India*

[2]*Department of Electronics and Communication Engineering, CMR Institute of Technology, Hyderabad, Telangana, India*

[3]*Department of Electronics and Communication Engineering, TRR Engineering College, Hyderabad, Telangana, India*

Corresponding author. E-mail: kpatteti@gmail.com

ABSTRACT

We studied the massive multiple-input multiple-output system performance with N-antenna users, as the advantage of N streams can be multiplexed per user, increasing the channel estimation overhead linearly with N. Spectral efficiency (SE) of uplink and downlink expressions are derived for any N-antenna user and these are achievable using estimated channels and per-user basis minimum mean-squared error successive interference cancellation (MMSE-SIC) detectors. This analysis shows that MMSE-SIC has similar asymptotic SE as linear MMSE detectors indicating that the SE increase from having multiantenna users can be harvested using linear detectors. Also we generalize the power scaling laws for massive MIMO to handle arbitrary N and show that one can reduce the multiplication of the pilot power and payload power as 1/M where M is the number of base station antennas, and still notably increase the SE with M before reaching a non-zero asymptotic limit. Simulations show that SE increase with N-antenna users, also note that the same improvement can be achieved by serving N times more single-antenna users instead. Thus the additional user antennas are particular useful for SE improvements when there are few active users in the system.

10.1   INTRODUCTION

One of the attractive huge research interests from last few years in wireless multiuser communication technologies is massive multiple-input multiple-output (MIMO) system. By utilizing hundreds of antennas at the base station (BS) and serving tens of users in each cell simultaneously, a drastic increase in SE can be achieved and simple coherent linear processing techniques.[1-3] Hence, massive MIMO is one of the key technologies for the next-generation wireless communication networks.

Literature of massive MIMO focuses only on single-antenna user terminals;[1-3] however, contemporary user terminals already feature multiple antennas to enhance the SE of the networks as well as the users.[4] Therefore, many devices, for example, laptops and vehicles, have moderate physical sizes; the deployment of 5 or 10 antennas per device is highly realistic, particularly for systems that operate at millimeter wave frequencies.[5] It is necessary to evaluate the performance analysis for massive MIMO systems with multi-antenna users, how the additional antennas should be useful for increasing the SE.

In addition, capacity analysis has been investigated for small-scale MIMO systems with multi-antenna users, but mainly with perfect channel state information[6,7] (CSI) and imperfect CSI in point-to-point and multiple access MIMO system[8-10] but no large system analysis is provided for the study of massive MIMO system behavior. A fixed CSI[11] analysis claimed that it is better to serve many single-antenna users than fewer multi-antenna users.

In this chapter, we analyze the SE of a massive MIMO system with estimated CSI and any number of antennas each per user. Lower bounds on the sum capacity are derived for uplink and downlink, which are achievable by per-user basis MMSE-SIC detectors and only uplink pilots. This analysis shows that users equipping with multiple antennas can greatly enhance the SE, particularly in lightly loaded systems where there are very few users to exploit the full multiplexing capability of massive MIMO with single antenna ($N = 1$) and the benefits can harvested by linear processing.

10.2   SYSTEM MODEL

We consider a single-cell system in time division duplex mode where the BS has M antennas and serves k users within each time-frequency coherence block. Each user is equipped with N antennas. We assume that each coherence block contains S transmission symbols and the channels of all users remain unchanged within each block.

Let $G_k \in \mathbb{C}^{M \times N}$ denote the channel response from user k to the BS within a coherence block. The fading can be spatially correlated, due to insufficient spacing between antennas and insufficient scattering in the channel. We use the classical Kronecker model to describe the spatial correlation[12]

$$G_k = R_{r,k}^{1/2} G_{w,k} R_{t,k}^{1/2} \tag{10.1}$$

where entries of $G_{w,k} \in \mathbb{C}^{M \times N}$ follow independent and identically distributed (i.i.d.) zero-mean circularly symmetric complex Gaussian distributions. $G_{t,k} \in \mathbb{C}^{N \times N}$ represents the spatial correlation at user k and $G_{r,k} \in \mathbb{C}^{M \times M}$ describes the spatial correlation at the BS for the link to user k. The large-scale fading parameter is included in $R_{r,k}$ and can be extracted as $(1/M) \, \mathrm{tr} \, (R_{r,k})$. Let $R_{r,k} = U_k \Lambda_k U_k^H$ be the eigenvalues decomposition of $R_{r,k}$ and $U_k \in \mathbb{C}^{N \times N}$ is a unitary matrix and $\Lambda_k = \mathrm{diag} \, \{\lambda_{k,1} \ldots, \lambda_{k,N}\}$ contains the eigenvalues.

10.2.1   UPLINK CHANNEL ESTIMATION

During the uplink pilot signaling, $B = Nk$ orthogonal pilot sequences are needed to estimate all channel dimensions at the BS. The pilot matrix of user k is $F_k \in \mathbb{C}^{N \times B}$. Suppose each user only knows its own statistical CSI; $R_{t,k}$ then based on Ref. [13] the pilot matrix that minimizes the mean square error of channel estimation under the pilot energy constraint $\mathrm{tr}\left(F_k F_k^H\right) \le BP_k$ has the form of $F_k = U_k L_k^{1/2} V_k^T$ where P_k is the maximum transmit power of user k, $L_k = \mathrm{diag} \, \{l_{k,1} \ldots, l_{k,N}\}$ distributes this power among the N channel dimensions, and $V_k \in \mathbb{C}^{B \times N}$ satisfies $V_k^H V_k = BI_N$ and $V_k^H V_l = 0$ if $k \ne l$. Thus, the received signal at BS is

$$Y = \sum_{k=1}^{K} G_k F_k + N = \sum_{k=1}^{K} H_k D_k^{1/2} V_k^T + N \in \mathbb{C}^{M \times B} \tag{10.2}$$

where we define $H_k = R_{r,k}^{1/2} G_{w,k} U_{t,k}$ and $D_k = \Lambda_k L_k$ with $d_{k,i}$ being its ith diagonal element. N is the receiver noise that follows vec $(N) \sim CN(0, \sigma^2 I_{BM})$ where vec($\cdot$) is the vectorization operator. Assume that the BS knows the statistical information D_k; then from Ref. [13], the MMSE estimate of $\hat{h}_k = \mathrm{vec}(H_k)$ is

$$\hat{h}_k = \left(D_k^{1/2} \otimes R_{r,k} \right)\left[\left(D_k \otimes R_{r,k} \right) + \frac{\sigma^2}{B} I_{MN} \right]^{-1} b_k \qquad (10.3)$$

where $b_k = \mathrm{vec}\left((1/B)Y_k V_k^*\right) = \mathrm{vec}\left(H_k D_k^{1/2} + \left(1/\sqrt{B}\right)NV_k^* \right)$ and $\otimes$ denotes the Kronecker product. Let $\hat{h}_{k,i}$ be the ith column of $\hat{H}_k$, then

$$E\left\{ \hat{h}_{k,i}\, \hat{h}_{k,j}^H \right\} = \begin{cases} \Phi_{k,i}, & i=j \\ 0, & i \neq j \end{cases} \qquad (10.4)$$

where

$$\Phi_{k,i} = d_{k,i} R_{r,k}\left(d_{k,i} + \frac{\sigma^2}{B} I_M \right)^{-1} R_{r,k}$$

10.2.2   UPLINK ACHIEVABLE SPECTRAL EFFICIENCY

When the receiving BS knows the perfect CSI of all users while each transmitter has only its own statistical CSI, the precoding directions of each user which maximize the sum capacity coincide with the eigenvectors of their own spatial correlation matrix.[14] Let $F_k \in \mathbb{C}^{N \times N}$ denote the precoding matrix of user k in the uplink payload data transmission phase, then $\tilde{F}_k = U_k P_k^{1/2}$, where $P_k = \mathrm{diag}\ \{p_{k1},....,p_{kN}\}$ with $(P_k) \leq p_k$ is the power allocation matrix. Although in our work the BS is only aware of the estimated CSI, $F_k = U_k P_k^{1/2}$ is still a reasonable option to enhance the SE. Hence, the received signal at the BS is

$$y = \sum_{k=1}^{K} G_k F_k x_k + n = \sum_{k=1}^{K} H_k \Lambda_k^{1/2} P_k^{1/2} x_k + n \qquad (10.5)$$

where x_k is the transmitted data symbol from user k and n is additive receiver noise. Since the BS is only aware of the estimated CSI, the effects of the channel uncertainty on the mutual information of MIMO channels need to be addressed. For our system and signal model, we develop a lower bound on the mutual information between $x = [x_1,....,x_k]$ and y in the following theorem.

Theorem 10.1: Consider the multiple access MIMO channel in (10.5), given imperfect CSI $\hat{H} = \left[\hat{H}_1, \ldots, \hat{H}_k\right]$ at the BS where $\hat{h}_k = \text{vec}(H_k)$ is given in (10.3). A lower bound on the mutual information between $x = [x_1, \ldots, x_k]$ and y is

$$I\left(y, \hat{H}; x\right) \geq \sum_{k=1}^{K} E\left\{\log_2 \left| I_N + Q_k \hat{H}_k^H \sum_k \hat{H}_k \right| \right\} \triangleq \sum_{k=1}^{K} R_{ul,k}^{SIC} \qquad (10.6)$$

where $Q_k = \Lambda_k P_k$ and $\sum_k = \left(\sum_{l \neq k} \hat{H}_l Q_l \hat{H}_l^H + Z + \sigma^2 I_M\right)^{-1}$ with

$$Z = \sum_{l=1}^{K} \sum_{n=1}^{N} \lambda_{l,n} p_{l,n} \left(R_{r,l} - \Phi_{l,n}\right)$$

The expectation is computed with respect to the channel estimates and $|\cdot|$ denotes the determinant of a matrix. Since the SIC procedure can be computationally complex, another option is to treat the N data streams as being transmitted by N independent users and use a linear MMSE detector to detect the NK streams independently. Based on the same methodology as in Ref. [15], the MMSE detector that maximizes the uplink SE of the *i*th stream of user k is

$$f_{k,i} = \sqrt{\lambda_{k,i} p_{k,i}} \sum \hat{h}_{k,i} \qquad (10.7)$$

where $\sum = \left(\sum_k^{-1} + \hat{H}_k Q_k \hat{H}_k^H\right)^{-1}$ applying the linear detector $f_{k,i}$ to the signal in (10.5); an uplink achievable SE of user *k* is

$$R_{ul,k}^{MMSE} = \sum_{i=1}^{N} E\left\{\log_2 \left(1 + \eta_{k,i}^{ul}\right)\right\} \qquad (10.8)$$

where the SINR of the *i*th stream is

$$\eta_{k,i}^{ul} = \frac{\lambda_{k,i} p_{k,i} \left| f_{k,i}^H \hat{H}_{k,i} \right|^2}{E\left\{ f_{k,i}^H \left(yy^H - \lambda_{k,i} p_{k,i} \hat{h}_{k,i} \hat{h}_{k,i}^H \right) f_{k,i} \middle| \hat{H} \right\}} \qquad (10.9)$$

Since interference from the user's own streams is not suppressed by $f_{k,i}$, it is intuitive that $R_{ul,k}^{SIC} \geq R_{ul,k}^{MMSE}$.

10.2.3   DOWNLINK ACHIEVABLE SPECTRAL EFFICIENCY

To limit the estimation overhead, we assume no downlink pilot or CSI feedback from the BS to users. This is common practice in massive MIMO since only the BS needs CSI to achieve channel hardening. Hence, the user has no instantaneous CSI except to learn the average effective channel $\hat{H}_k \triangleq \Lambda_k^{1/2} E\{H_k^H W_k\} \Omega_l^{1/2}$ and covariance matrix of the interference term.

Let $W_k \in \mathbb{C}^{M \times N}$ be the downlink precoding matrix associated with user k and let $\Omega_k = \mathrm{diag}\{w_{k,1},\ldots,w_{k,N}\}$ allocate the total transmit power P_k' among the N streams. Then, the total transmit power from the BS is $\sum_{k=1}^{K} P_k'$. The received signal at user k is

$$y_k = G_k^H \sum_{l=1}^{K} W_l \Omega_l^{1/2} x_l + n_k \in \mathbb{C}^{N \times 1} \qquad (10.10)$$

where x_l the downlink signal is intended for user l and n_k is the additive receiver noise. Without loss of generality, let user k use U_k^H (the eigenvector matrix of its own correlation matrix) as a first step detector to adapt to the channel correlation; then, the processed received signal is

$$z_k = U_k^H y_k = \Lambda_k^{1/2} H_k^H \sum_{l=1}^{K} W_l \Omega_l^{1/2} x_l + U_k^H n_k \qquad (10.11)$$

A lower bound on the mutual information $I(z_k;x_k)$ is developed in the following theorem.

Theorem 10.2: Consider the downlink signal model in (10.11), given the average effective channel $\bar{H}_k \triangleq \Lambda_k^{1/2} E\{H_k^H W_k\} \Omega_l^{1/2}$ of user k. The mutual information between z_k and x_k is

$$I(z_k;x_k) \geq \log_2 \left| I_N + \bar{H}_k^H \Xi_k \hat{H}_k \right| \triangleq R_{dl,k}^{SIC} \qquad (10.12)$$

where $\Xi_k = \left(\Lambda_k^{1/2} E\left\{ H_k^H \sum_{l \neq k} (W_l \Omega_l W_l^H) H_k \right\} \Lambda_k^{1/2} + \sigma^2 I_N \right)^{-1}$

The lower bound in Theorem 10.2 can be achieved if user k applies MMSE-SIC detection to z_k when regarding $\bar{H}_k$ as the true channel and the uncorrelated term $z_k - \bar{H}_k x_k$ is treated as worst-case Gaussian noise in the detector. Theorem 10.2 generalizes the conventional SE analysis of massive MIMO from $N = 1$ to arbitrary N.

The user can also apply a linear MMSE detector for symbol detection based on (10.11). Denote $\overline{h}_{k,i}$ as the ith column of $\hat{H}_k$, and then with the knowledge of $\overline{H}_k$, the MMSE detector for the ith stream of user k that maximizes the corresponding downlink SE is $r_{k,i} = \Xi_k \overline{h}_{k,i}$ where $\Xi_k = \overline{\Xi}_k^{-1} + \overline{H}_k \overline{H}_k^H$. Applying $r_{k,i}$ to (10.11), the achievable SE of user k is

$$R_{dl,k}^{MMSE} = \sum_{i=1}^{N} E\left\{ \log_2\left(1 + \eta_{k,i}^{dl}\right) \right\} \tag{10.13}$$

where the SINR $\eta_{k,i}^{dl}$ of its ith stream is

$$\eta_{k,i}^{dl} = \frac{\left| r_{k,i}^H \overline{h}_{k,i} \right|^2}{r_{k,i}^r E\left\{ z_k z_k^H \right\} r_{k,i} - \left| r_{k,i}^H \overline{h}_{k,i} \right|^2} \tag{10.14}$$

Intuitively, the MMSE-SIC detector will have a higher performance than the MMSE detector in the downlink.

10.3   ASYMPTOTIC ANALYSIS

In this section, approximations of the SEs in Theorems 10.1 and 10.2 that are tight for large systems are derived for fixed power matrices L_k, P_k, and Ω_k. We consider the large system regime where M and k go to infinity while N remains constant since the users are expected to have a relatively small number of antennas.

Theorem 10.3: For the uplink MMSE-SIC detector on a per-user basis, a large-system approximation of $R_{ul,k}^{SIC}$ in Theorem 10.1 is

$$\overline{R}_{ul,k}^{SIC} \triangleq \sum_{i=1}^{N} \log_2\left(1 + \frac{1}{M}\text{tr}\left(\Phi_{k,i} T\right) \lambda_{k,i} p_{k,i}\right) \tag{10.15}$$

such that $R_{ul,k}^{SIC} - \overline{R}_{ul,k}^{SIC} \xrightarrow[M \to \infty]{} 0$, where $T = T\left(\sigma^2/M\right)$. In comparison, the large-system SE approximation of the linear MMSE detector $f_{k,i}$ can be derived by following the same procedures in Ref. [15].

The SE approximation is $\overline{R}_{ul,k}^{MMSE} = \sum_{i=1}^{N} \log_2\left(1 + \overline{\eta}_{k,i}^{ul}\right)$, where

$$\overline{\eta}_{k,i}^{ul} = \frac{\lambda_{k,i} p_{k,i} \delta_{k,i}^2}{\displaystyle\sum_{(l,n)\neq(l,i)} \lambda_{l,n} p_{l,n} \frac{1}{M}\mu_{k,i,l,n} + \frac{1}{M}\vartheta_{k,i}} \tag{10.16}$$

where $\delta_{k,i} = (1/M)\operatorname{tr}(\Phi_{k,i}T)$, $\mu_{k,i,l,n} = \operatorname{tr}(\Phi_{l,n}T_{k,i})/M(1+\lambda_{q,n}p_{l,n}\delta_{l,n})^2$, and $\vartheta_{k,i} = (1/M)\operatorname{tr}(\Phi_{k,i}T'')$.

By comparing (10.16) and Theorem 10.1, we can see that for the MMSE-SIC detector, the interstream interference of a user caused by imperfect CSI vanishes asymptotically, and only the interuser interference remains. For the linear MMSE detector, however, the interstream interference $\mu_{k,i,l,n}/M$ remains in (10.16) as well. However, the impact of this part reduces to zero as M grows. It shows that the SE improvements with multi-antenna users can be harvested in massive MIMO by linear detectors; thus, a simple hardware implementation is possible. Next, we derive large-system approximations of the downlink performance. The precoder used by the BS can be any linear precoder such as the matched filtering (MF), block diagonal zero-forcing, or MMSE precoding. Due to the limited space, we only consider the MF case

$$W_k = \frac{1}{\sqrt{E\left\{\operatorname{tr}\left(\hat{H}_k\hat{H}_k^H\right)\right\}}}\hat{H}_k \tag{10.17}$$

Theorem 10.4: For the downlink MMSE-SIC detector and the linear MMSE detector, if the BS utilizes the MF precoder, the large-system approximations of the SEs in Theorem 10.2 and (10.14) are the same, that is,

$$R_{dl,k}^{\mathrm{SIC}} \triangleq \sum_{i=1}^{N}\log_2\left(1+\frac{\lambda_{k,i}\omega_{k,i}\left(\alpha_{k,i}^2/\theta_k\right)}{(1/M)\gamma_k\lambda_{k,i}+\left(\sigma^2/M\right)}\right) \tag{10.18}$$

such that $R_{dl,k}^{\mathrm{SIC}} - \overline{R}_{dl,k}^{\mathrm{SIC}} \xrightarrow[M\to\infty]{} 0$, where $\theta_k = \sum_{i=1}^{N}\alpha_{k,i}$, $\alpha_{k,i} = (1/M)\operatorname{tr}(\Phi_{k,i})$, and $\gamma_k = (1/M)\operatorname{tr}\left(R_{r,k}\sum_{l\neq k}\sum_{i=1}^{N}(\omega_{l,i}/\theta_l)\Phi_{l,i}\right)$.

Theorem 10.4 shows that the SIC processing at users does not bring any advantage over the linear MMSE detector[17] in the downlink. The reason is that $\overline{H}_k$ is a diagonal matrix, which means that no interstream interference is introduced in this assumed true channel. Therefore, the SIC processing is neither necessary nor beneficial when there are no uplink pilots. This result has positive influence on the design of user devices since it indicates low hardware requirements and simplifies the SE optimization.

10.4   POWER SCALING LAWS

It is shown in Refs. [2,3] that for $N = 1$, the transmit power can be reduced with retained performance as the number of BS antennas grows. Next, we generalize the fundamental result to handle any fixed N. Assume the pilot power is reduced as $L_k = \left(1/M^\alpha\right)L_k^{(0)}$ and the payload powers are $P_k = \left(1/M^{1-\alpha}\right)P_k^{(0)}$ and $\Omega_k = \left(1/M^{1-\alpha}\right)\Omega_k^{(0)}$ where $0 \le \alpha \le 1$ and the $(\cdot)^{(0)^k}$ matrices are fixed. We consider $R_{r,k} = \beta_k I_M$ where β_k is the large-system fading of user k so that the correlation matrix at the BS remains unchanged as M grows. A different large-system limit is considered in this section: M goes to infinity while K and N are fixed.

For the uplink MMSE-SIC receiver on a per-user basis, if the pilot power is reduced as $L_k = \left(1/M^\alpha\right)L_k^{(0)}$ and the payload power is $P_k = \left(1/M^{1-\alpha}\right)P_k^{(0)}$, then $R_{ul,k}^{\text{SIC}} - \overline{R}'_{ul,k} \underset{M\to\infty}{\to} 0$, where

$$\overline{R}'_{ul,k} = \sum_{i=1}^{N} \log_2\left(1 + \beta_k^2 \lambda_{k,i}^2 \frac{Bl_{k,i}^{(0)} p_{k,i}^{(0)}}{\sigma^2\left(z + \sigma^2\right)}\right) \tag{10.19}$$

where $z = 0$, if $0 \le \alpha < 1$ and $z = \sum_{l=1}^{K} \beta_l \text{tr}\left(\Lambda_l P_l^{(0)}\right)$ if $\alpha = 1$.

Lemma 10.1: For the downlink MMSE-SIC detector and the MMSE detector, if $L_k = \left(1/M^\alpha\right)L_k^{(0)}$ and the payload power is $\Omega_k = \left(1/M^{1-\alpha}\right)\Omega_k^{(0)}$, then $R_{dl,k}^{\text{SIC}} - \overline{R}'_{dl,k} \underset{M\to\infty}{\to} 0$, where

$$\overline{R}'_{dl,k} = \sum_{i=1}^{N} \log_2\left(1 + B\beta_k^2 \lambda_{k,i}^2 \frac{\omega_{k,i}^{(0)} l_{k,i}^{(0)} \upsilon_{k,i}}{\sigma^2\left(\beta_k \lambda_{k,i} \gamma + \sigma^2\right)}\right) \tag{10.20}$$

where $\upsilon_{k,i} = \lambda_{k,i} l_{k,i}^{(0)} / \text{tr}\left(\Lambda_k L_k^{(0)}\right) \in [0,1]$, $\gamma = 0$ if $0 \le \alpha \le 1$ and $\gamma = \sum_{l=1}^{K}\sum_{i=1}^{N} \omega_{l,i}^{(0)} \upsilon_{l,i}$ if $\alpha = 1$.

Notice that $\overline{R}'_{ul,k}$ and $\overline{R}'_{dl,k}$ are fixed nonzero values independent of M. Consequently, when the number of BS antennas is large enough, we can reduce the multiplication of the pilot power and the payload power as $1/M$ and achieve a nonzero asymptotic fixed SE. When $\alpha = 0.5$ and $N = 1$, our results reduce to the $1/\sqrt{M}$ scaling law for the pilot/payload powers proposed by Ref. [2].

10.5   SIMULATIONS RESULTS

We consider a cell with a radius of 500 m. The user locations are uniformly distributed at distances to the BS of at least 70 m. Statistical channel inversion power control is applied in the uplink, equal power allocation is used in the downlink, and the power is divided equally between the N streams of each user; that is, $\beta_l l_{l,i} = \beta_l p_{l,i} = p/N\sigma^2$ and $\omega_{l,i} = P_d$ where $\beta_l = (1/M)\operatorname{tr}(R_{r,l})$ with ρ/σ^2 being set to 0 dB. P_d is set to a value such that the cell-edge SNR (without shadowing) is −3 dB. The exponential correlation model from Ref. [16] is used for $R_{t,k}$ and $R_{r,k}$. The correlation coefficients between adjacent antennas at the BS and at the users are $a_r e^{j\theta_{r,k}}$ and $a_t e^{j\theta_{t,k}}$, respectively, with $a_r = a_t = 0.4$ and $\theta_{r,k}$, $\theta_{t,k}$ uniformly distributed in $[0, 2\pi]$. The coherence block length is $S = 200$, which supports high user mobility.

The uplink and downlink sum SE of the MMSE-SIC and MMSE detectors are shown in Figure 10.1. It shows that the two detectors achieve almost the same SEs, which verifies the conclusion that a linear detector can achieve most of the SE improvements from equipping users with multiple antennas in massive MIMO. Moreover, although the pilot overhead increases, 90% and 75% performance gains are achieved for the uplink and the downlink, respectively, by increasing N from 1 to 3 for $M = 200$. Figure 10.1 also verifies the tightness of the large-system approximations derived in Theorems 10.3 and 10.4.

Figure 10.2 confirms the power scaling laws in Lemmas 1 and 2. Results for $\alpha = 0.5$ and $\alpha = 1$ are shown. It is observed that, even with a $1/M$ reduction of the multiplication of pilot and payload powers, a notable increase of SE can still be obtained for an extremely wide range of M before reaching the limit, especially for $M \in [50, 1000]$ which is of practical interest.

Recall that the channel estimation overhead Nk equals the number of data streams that are transmitted. For a fixed number of data streams Nk, the system can schedule Nk single-antenna users and send one stream to each user, or schedule fewer multi-antenna users and send several streams to each. The downlink performance of these different scheduling approaches is compared in Figure 10.3 for $N \in \{1, 3, \text{and } 10\}$.

The power per stream is P_d as Figures 10.1 and 10.3 show that for any given Nk, scheduling Nk single-antenna users is always (slightly) beneficial.

The optimal Nk is around 100, which requires 100 active users per coherence block if $N = 1$. With multi-antenna users, more realistic user numbers are sufficient to reach the sweet spot of $Nk \approx 100$. Therefore, additional user antennas are beneficial to increase the spatial multiplexing in light and medium-loaded systems.

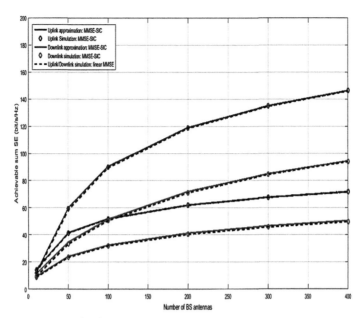

FIGURE 10.1   (See color insert.) Uplink and downlink achievable sum SE as a function of the number of BS antennas for $K = 10$.

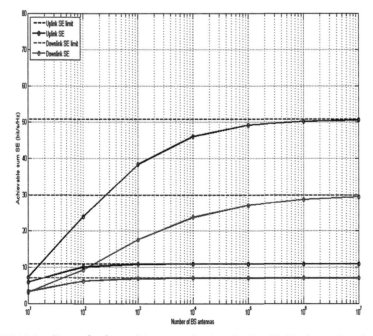

FIGURE 10.2   (See color insert.) Power scaling law for $K = 10$, $N = 3$, $a_r = 0$, and $a_t = 0.4$.

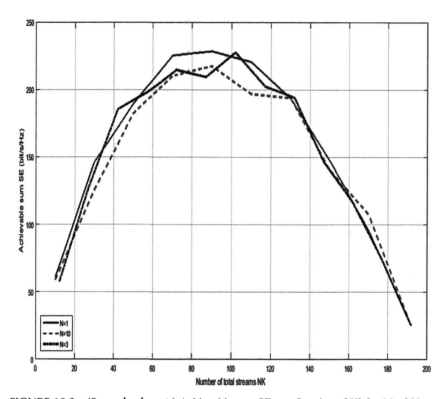

FIGURE 10.3 **(See color insert.)** Achievable sum SE as a function of Nk for $M = 200$.

10.6   CONCLUSIONS

We analyzed the achievable SE of single-cell massive MIMO systems with multi-antenna users. With estimated CSI from uplink pilots, lower bounds on the ergodic sum capacity were derived for both the uplink and the downlink, which are achievable by per-user MMSE-SIC detectors. Large-system SE approximations were derived and show that the MMSE-SIC detector has an asymptotic performance similar to the linear MMSE detector, indicating that linear detectors are sufficient to handle multi-antenna users in massive MIMO. We generalized the power scaling laws for massive MIMO from $N = 1$ to arbitrary N. We showed that the SE increases with N, but for a fixed value of Nk, the highest SE is achieved by having Nk single-antenna users. Hence, additional user antennas are mainly beneficial to increase the spatial multiplexing in systems with few users.

KEYWORDS

- **CSI**
- **SE**
- **MMSE-SIC**
- **power scaling**
- **massive MIMO**

REFERENCES

1. Marzetta, T. L. Noncooperative Cellular Wireless with Unlimited Numbers of Base Station Antennas. *IEEE Trans. Wireless Commun.* **2010,** *9* (1), 3590–3600.

2. Ngo, H. Q.; Larsson, E. G.; Marzetta, T. L. Energy and Spectral Efficiency of Very Large Multiuser MIMO Systems. *IEEE Trans. Commun.* **2013,** *61* (4), 1436–1449.

3. Hoydis, J.; ten Brink, S.; Debbah, M. Massive MIMO in the UL/DL of Cellular Networks: How Many Antennas Do We Need? *IEEE J. Sel. Areas Commun.* **2013,** *31* (2), 160–171.

4. Gesbert, D.; Kountouris, M.; Heath, R. W.; Chae, C.-B.; Salzer, T. Shifting the MIMO Paradigm. *IEEE Trans. Signal. Process* **2007,** *24* (5), 36–46.

5. Swindlehurst, A. L.; Ayanoglu, E.; Heydari, P.; Capolino, F. Millimeter-Wave Massive MIMO: The Next Wireless Revolution? *IEEE Commun. Mag.* **2014,** *52* (9), 56–62.

6. Jindal, N.; Goldsmith, A. Dirty-Paper Coding versus TDMA for MIMO Broadcast Channels. *IEEE Trans. Inf. Theory* **2005,** *51* (5), 1783–1794.

7. Taesang, Y.; Goldsmith, A. On the Optimality of Multiantenna Broadcast Scheduling Using Zero-Forcing Beamforming. *IEEE J. Sel. Areas Commun.* **2006,** *24* (3), 528–541.

8. Yoo, T.; Goldsmith, A. Capacity and Power Allocation for Fading MIMO Channels with Channel Estimation Error. *IEEE Trans. Inf. Theory* **2006,** *52* (5), 2203–2214.

9. Layec, P.; Piantanida, P.; Visoz, R.; Berthet, A. O. Capacity Bounds for MIMO Multiple Access Channel with Imperfect Channel State Information. *Proc. IEEE ITW* **2008,** *4,* 21–25.

10. Musavian, L.; Nakhai, M. R.; Dohler, M.; Aghvami, A. H. Effect of Channel Uncertainty on the Mutual Information of MIMO Fading Channels. *IEEE Trans. Vehicul. Tech.* **2007,** *56* (5), 2798–2806.

11. Björnson, E.; Kountouris, M.; Bengtsson, M.; Ottersten, B. Receive Combining vs. Multi-stream Multiplexing in Downlink Systems with Multi-antenna Users. *IEEE Trans. Signal Process.* **2013,** *61* (13), 3431–3446.

12. Kermoal, J. P.; Schumacher, L.; Pedersen, K. I.; Mogensen, P. E.; Frederiksen, F. A Stochastic MIMO Radio Channel Model with Experimental Validation. *IEEE J. Sel. Areas Commun.* **2002,** *20* (6), 1211–1226.

13. Björson, E.; Ottersten, B. A Framework for Training-Based Estimation in Arbitrarily Correlated Rician MIMO Channels with Rician Disturbance. *IEEE Trans. Signal Process.* **2010,** *58* (3), 1807–1820.

14. Li, X.; Gao, X. Q.; McKay, M. R. Capacity Bounds and Low Complexity Transceiver Design for Double-Scattering MIMO Multiple Access Channels. *IEEE Trans. Signal Process* **2010,** *58* (5), 2809–2822.

15. Li, X.; Björnson, E.; Larsson, E. G.; Zhou, S.; Wang, J. A Multicell MMSE Detector for Massive MIMO Systems and New Large System Analysis. In *Proc. IEEE GLOBECOM,* Dec. 2015.

16. Loyka, S. Channel Capacity of MIMO Architecture Using the Exponential Correlation Matrix. *IEEE Commun. Lett.* **2001,** *5* (9), 369–371.

17. Wagner, S.; Couillet, R.; Debbah, M.; Slock, D. T. M. Large System Analysis of Linear Precoding in Correlated MISO Broadcast Channels under Limited Feedback. *IEEE Trans. Inf. Theory* **2012,** *58* (7), 4509–4537.

CHAPTER 11

NEIGHBOR NODES DISCOVERY SCHEMES IN A WIRELESS SENSOR NETWORK: A COMPARATIVE PERFORMANCE STUDY

SAGAR MEKALA[1,2] and K. SHAHU CHATRAPATI[3*]

[1]JNTUH, Hyderabad, India

[2]Department of CSE, UCET, Mahatma Gandhi University, Nalgonda, Telangana, India

[3]Department of CSE, JNTUH College of Engineering Manthani, Karimnagar, Telangana, India

*Corresponding author. E-mail: shahujntu@jntuh.ac.in

ABSTRACT

Technical advances in embedded systems have prompted an expansion in the quantity of little-estimated detecting and imparting gadgets outfitted with remote interfaces. Systems of such gadgets discover relevance in observing and reconnaissance exercises for wireless sensor networks (WSNs) vehicular activity policing, human informal organizations, and so forth. It is, for the most part, anticipated that the quantity of gadgets will increment amid the following decades, prompting an ascent in the quantity of utilization situations including expensive systems. Especially, neighbor revelation is a principal building hinder for WSNs, on the grounds that it is the initial step to set up correspondence connects between sensor hubs. Customary neighbor disclosure issues, for the most part, concentrate on static remote systems where all hubs work on the same recurrence. In any case, the multiplication of cell phones and multichannel correspondences present new difficulties on this issue. In this chapter, we have performed an extensive comparative performance analysis of various neighbor discovery schemes in WSN.

11.1   INTRODUCTION

Contemporary changes in wireless foundations and microelectronics have bolstered the augmentation of one spending plan, little control, multi-operational sensor hubs, those are in little size, and the correspondence is made in little range. These hubs having an ability of detecting, preparing for information, and conveying segments control the possibility of sensor systems. Sensor systems are shaped utilizing gathering of nonconcurrent or synchronous hubs. These sensors will focus the association with each other in wireless sensor network (WSN) by cross-region structure.[1]

The sensors in WSN[2] can sense differing occasions delicately. Some sensors focus about the changes and pass the message or the record starting with one specific sensor focus point, then onto the accompanying sensor focus. The wide applications are incorporated with both nonmilitary work force and military circumstances, including regular checking, observation for prosperity and security, robotized restorative administrations, sharp building control, and development control. With the change of PC gear advancement, the CPU and blast memory are getting the opportunity to be tinier, smaller, more exceptional, and less costly (Fig. 11.1).[3]

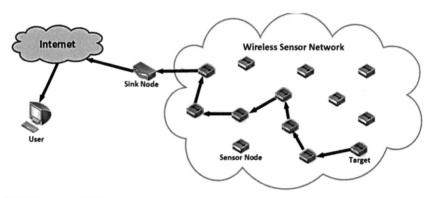

FIGURE 11.1   WSN architecture.

Subsequently, the memory and get-ready capacities of sensor centers won't be the most basic obstruction for the usage of WSNs. In any case, the battery development may fail to get a jump forward. WSN has trans-formed its characteristics due to key bottlenecks.[4] So, the investigation on essentialness viability of WSNs is still the center premium. Fundamentally, WSNs directing neighbor disclosure is a noteworthy research movement.

Beforehand, researchers have concentrated on the neighbor revelation utilizing Hi-parcel transmission. In any case, the hubs with low power are not investigated in the system but rather still it is a noteworthy issue. For the wide suitability extent of WSNs, it is hard to manufacture WSN-guiding figuring that fulfills all application necessities or maybe it is of noteworthiness that arranging general coordinating figuring which by some methods can be associated with a couple of uses and in the meantime modify the essentialness usage to construct the framework lifetime very far. At this moment, there are amazing plans of investigation and tries that are on the go, for the change of controlling traditions in WSNs. The sensors ought to endlessly scan for new neighbors to oblige the going with conditions[5]:

1. Loss of adjacent synchronization due to totaled clock coasts.
2. Aggravation of remote accessibility between bordering centers by a break event, for instance, a passing auto or animal, a tidy storm, rain, or cloudiness; when these events are over, the disguised center points must be rediscovered.
3. The advancing development of new center points, in a few frameworks, to compensate for centers that have ceased to deal with the grounds that their essentialness has been exhausted.
4. The development in transmission compels of a couple of center points, in light of particular events, for instance, the disclosure of new conditions.

Especially vitality primary imperative in neighbor disclosure, because of the absence of vitality, the hubs can't investigate utilizing Hi-parcel transmission; yet, the hubs still stow away in the system and cause the system at awkwardness state.[6] In this chapter, we investigate about different mechanisms of existing neighbor discovery. Rest of the chapter is organized as follows: Section 11.2 describes the different methodologies of neighbor discoveries, Section 11.3 illustrates the comparative analysis of the different mechanisms, and Section 11.4 concludes the chapter.

11.2 NEIGHBOR-DISCOVERY MECHANISMS

You et al.[7] proposed "Salud Like Neighbor Discovery in Low-Duty-Cycle Wireless Sensor Networks" the issue of neighbor disclosure when hubs utilize (simple random access *protocol)* ALOHA-like opened revelation calculations in low-obligation cycle WSNs. This issue is noninconsequential on the grounds that a hub may need to transmit commonly without crashes to make every one of its neighbors find it, while one time is sufficient for all-hub dynamic systems. By diminishing the examination to K coupon collector's problem, authors demonstrate that the normal time to find all its $n - 1$ neighbors for every hub is upper limited by $ne(\log_2 n + (3 \log_2 n - 1) \log_2 n + c)$ for some steady c with high likelihood, and it is lower limited by $ne\ln n + cn$, where c is a positive consistent and e is the base of regular logarithm. In addition, the authors demonstrate that the disclosure time is around the desire. Likewise, in this scheme, the authors extend the ALOHA-like calculation to manage the situation when hubs don't have a clue about the quantity of neighbors. It just prompts at most an element of two stop pages for upper bound contrasted and known number of neighbors. At that point, the authors accept the hypothetical results by broad reenactments and investigate the diverse calculation execution in low-duty-cycle and nonobligation cycle WSNs. Additionally, the authors apply the way to deal with examine new situation of inconsistent connections in low duty-cycle WSNs. To the best of insight, it is the own work to investigate the execution of ALOHA-like neighbor revelation calculations in low-obligation cycle WSNs.

Khanmirza et al.[8] proposed "Assessing Passive Neighborhood Discovery for Low Power Listening MAC Protocols." In this chapter, the authors examine the likelihood of utilizing a totally aloof, low-power and zero-overhead neighbor revelation technique, particularly working in conjunction with offbeat LPL MACs. The key point behind the inactive detecting procedure is to profit by the data acquired from disentangling of the parcels amid clear channel assessment (CCA) checks. This infers definitely no cost for sensor hubs, as intermittent CCA checks and translating of the got bundles if there should be an occurrence of vitality location inside the channel are standard undertakings that all hubs ought to perform when they wake-up. Moreover, the authors contend that coordinating neighbor disclosure convention to LPL MAC layer rearranges the entire instrument fundamentally, as well as serves to a more intelligent lessening of force utilization.

Iyer et al.[9] proposed "NetDetect: Neighborhood Discovery in Wireless Networks Using Adaptive Beacons." In this chapter, the authors mainly concentrate on the issue of effectively finding neighbors in a remote system of hubs. Considering the class of conventions that depend on probabilistic transmissions which concentrate on expanding the quantity of disclosures per time unit, and search for procedures that permit hubs to find their neighbors as quick and as proficient as could be expected under the circumstances.

Existing methodologies target for the most part static, completely associated arranges and experience the ill effects of a bootstrap issue as all hubs need to begin at the same minute in time. The fundamental effect of the chapter is the plan and execution of a versatile and decentralized technique, named NetDetect that tackle the nearby neighborhood disclosure issue by misusing the beaconing instrument. The approach is to have hubs appraise the nearby neighborhood measure from the quantity of blunders distinguished on the correspondence channel with a most extreme probability estimator. The hub connection prompts a self-versatile instrument, where the beaconing likelihood quickly meets to the craved ideal. The calculation on an assortment of systems with expanding levels of progression: a completely associated system, static and portable multibounce works arrange. NetDetect performs well in every single considered situation, keeping up a high rate of neighbor disclosures and great appraisals of the area densities even in extremely dynamic circumstances. Correlation with existing methodologies demonstrates that the NetDetect plan is effective from both the union time and vitality viewpoint. In this chapter, the authors make the accompanying commitments[9]:

1. Distinguish a procedure of amplifying disclosure productivity organized appropriately (versatile),
2. outline and actualize NetDetect, a conveyed calculation wherein hubs adjust their probabilities of transmission in view of privately measured throughput, and
3. assess NetDetect in contrast with existing neighborhood revelation conventions and showcase its proficiency and vigor.

Authors utilized the mistakes on correspondence channel as the principle wellspring of data for assessing the nearby neighborhoods, as these data are constantly present at each hub in a remote correspondence environment. This prompts an exquisite arrangement just misusing effectively existing data. Measuring diverts dispute with a specific end goal to change the transmission channel recurrence or adjust conflict window sizes has as of now been looked into. All things considered, to the best of our insight, this is the principal versatile calculation misusing channel blunders data for inferring data in regards to the accessible neighbors, working in a dispersed situation, and having the capacity to track changes without extraclient activities. While we outlined NetDetect particularly for low power remote systems, the calculation is specifically material to different frameworks that can be demonstrated utilizing Poisson forms and which need a self-versatile conduct. The lightweight correspondence overhead of the calculation consolidated with its appropriateness on an expansive scope of situations portrayed by various system topologies and its heartiness, making NetDetect an alluring arrangement.

Kohvakka et al.[10] proposed "Energy-efficient neighbor discovery protocol for mobile wireless sensor networks." In this chapter, the authors present a new energy-efficient neighbor discovery protocol (ENDP) for synchronized low duty-cycle medium access control (MAC) schemes. The presented protocol reduces the need for network scans by distributing synchronization information from nodes in two-hop neighborhood. This information is carried in the beacon payloads of underlying MAC protocol and utilized for establishing new communication links. In addition, ENDP introduces an efficient network beacon signaling scheme to make network scans more energy efficient. ENDP is the first protocol that can effectively minimize network energy consumption in dynamic WSNs. The energy efficiency and operation fidelity are verified by analytical performance models and experimental measurements using real WSN prototypes.

Chen et al.[11] proposed "On Heterogeneous Neighbor Discovery in Wireless Sensor Networks;" in which the authors introduce two ideal neighbor disclosure conventions, called Hedis (heterogeneous discovery as a majority-based convention) and Todis (triple-odd-based discovery as a coprimality-based convention), that certification offbeat neighbor revelation in a heterogeneous domain, implying that every hub could work at an alternate obligation cycle. In particular, they streamline the obligation cycle granularity in their separate convention classes to bolster obligation cycles as 2n and 3n individually, where n is the whole number that accomplishes all obligation cycles littler than one. Scientifically, contrast these two conventions and existing state-of-the-workmanship conventions to affirm their optimality in the backing of obligation cycles, furthermore, think about them against each different as an examination between the two general classes of neighbor disclosure conventions (majority versus coprimality-based conventions). Our outcomes demonstrate that while the disclosure latencies are comparative for both conventions, Hedis as an ideal majority-based convention matches real obligation cycles a great deal more intently than Todis as a coprime-based convention.

Cohen et al.[1] proposed "Persistent Neighbor Discovery in Asynchronous Sensor Networks." The fundamental thought behind the nonstop neighbor disclosure plot is it proposes the undertaking of finding another hub which is separated among every one of the hubs that can distinguish. These hubs are portrayed as (1) likewise neighbors; (2) place with an associated section of hubs that have officially distinguished each other; and (3) hub additionally has a place with this fragment. This variable demonstrates the in-portion level of a concealed neighbor. With a specific end goal to exploit the proposed revelation conspire, the hub must gauge the estimation.

Sun et al.[12] proposed "Neighbor Discovery in Low-Duty-Cycle WSNs with Multi-Packet Reception." In this chapter, the issue of ND in low-duty-cycle WSNs with k-MPR radios was contemplated and directed inside and out of execution examination on ALOHA-like ND conventions with different expansions. The commitments in this chapter are recorded as takes after the following:

- First, to the best of our insight, first to consider the issue of ND utilizing MPR radios as a part of low-duty-cycle WSNs. Authors mainly demonstrate that MPR can fundamentally quicken the ND procedure, and accordingly, the length of ND in low-obligation cycle systems can be massively abbreviated. Here, mainly think about the ALOHA-like convention in k-MPR that organizes and demonstrates that the normal time required is $O(n\log n \log nk)$, where n is the faction measure, by diminishing the issue to a summed up type of K coupon collector's problem.
- Furthermore, when a criticism component is brought into the framework, the method demonstrates that it gives a $\log n$ change over the ALOHA-like convention.
- Finally, the method extends to the situation where the inner circle measure n is obscure and demonstrates that it brings about a component of two-log jam.

Huang et al.[13] proposed "EasiND: Effective Neighbor Discovery Algorithms for Asynchronous and Asymmetric-Duty-Cycle Multi-Channel Mobile WSNs" that present an effective neighbor discovery system named EasiND for asynchronous and asymmetrical duty-cycle multichannel mobile WSNs. First, the authors propose an optimal synchronous multichannel neighbor discovery algorithm based on quorum system, which can bind the discovery latency in a multichannel scenario with low power consumption. Second, the authors design an asynchronous neighbor discovery quorum system for multichannel WSNs. Theoretical analyses demonstrate that EasiND achieved 33.3% and 50% reduction in power-latency product when compared to U-Connect and Acc, respectively. Third, to enable EasiND to be applied to asymmetrical duty-cycle system, this method proposes an on-demand time slot activation scheme that combines random and cooperative methods together, which effectively reduces discovery latency. Finally, this method presents a channel scanning acceleration approach based on spatial frequency characteristics of discovered neighbors, which further decreases discovery latency.

11.3   PERFORMANCE ANALYSIS

In this section, we evaluate the discovery latency and network initialization time of

1.   ALOHA-like neighbor discovery in low duty cycle (ALNDLDC)[7]
2.   Evaluating passive neighborhood discovery for low power listening MAC protocols (PNDLPL)[8]
3.   NetDetect: Neighborhood discovery in wireless networks using adaptive beacons[9]
4.   ENDP for mobile wireless sensor networks (EEND)[10]
5.   Heterogeneous neighbor discovery in wireless sensor networks (HND)[11]
6.   Continuous neighbor discovery in asynchronous sensor networks (CND)[1]
7.   Neighbor discovery in low-duty-cycle wireless sensor networks with multipacket reception (NDLDC)[12]
8.   EasiND: Effective neighbor discovery algorithms for asynchronous and asymmetric-duty-cycle multichannel mobile WSNs[13]

While assessing the execution of above neighbor disclosure calculation, we concentrate on the accompanying two criteria: network initialization cost and latency. Network initialization cost is the time taken to form a network by a method.

Revelation inactivity: The aggregate slipped by the time that all sensors in the system spend amid the neighbor disclosure handle. Reproduction model design parameters of the reenactment are given in Table 11.1. In this reproduction environment, 50 sensor gadgets are haphazardly sent inside a field of 100 × 100 m. A MAC convention for recreation concentrates on takes after a Carrier-sense multiple access with collision avoidance (CSMA/CA) way. We set up the CC2420 radio as the radio module for correspondence. The CC2420 module is ordinarily utilized as a part of various genuine sensors arrange applications. The length of every space in a disclosure calendar is 15 ms. Figure 11.2 shows the number of neighbor's explored comparison for each method in a particular time interval. Figure 11.3 shows the network initialization time of each method.[2]

TABLE 11.1   Comparative Analysis of Existing Neighbor Node Discovery.

S. no.	Method	Percentage of latency done by a node to find its neighbors (%)	N/W initialization time in s (avg. no of nodes = 100)
1	ALNDLDC	28	200
2	PNDLPL-MAC Protocols	28.7	221
3	NetDetect	26	192
4	EEND-Mobile WSN	28	200
5	HND-Wireless Sensor Networks	29	211
6	Persistent-CND Asynchronous Sensor Networks	26	187
7	NDLDC-WSN with Multipacket Reception	27	202
8	EasiND	24	180

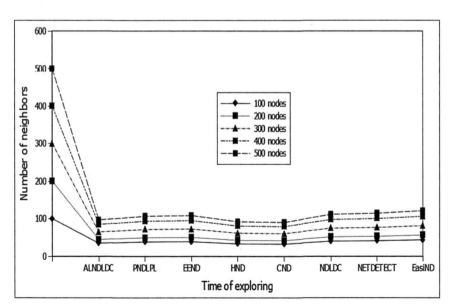

FIGURE 11.2   Latency analysis of neighbor nodes discovery schemes based on network size.

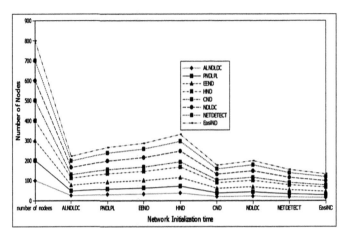

FIGURE 11.3 Comparison of network initialization cost of neighbor nodes discovery schemes based on network size.

11.4 CONCLUSION

We have evaluated the performance of various existing neighbor discovery algorithms in wireless sensor networks with the help of two performance metrics as neighbor nodes discovery per unit time for each method and network initialization cost in terms of network resources. Based on these performances metric, the existing schemes are evaluated and observed their performance deviations. The performance evaluation proved that EasiND scheme has achieved low latency to explore all the neighbors with low network cost consumption than other existing schemes. Hence, EasiND is better scheme than another scheme for neighbor nodes discovery in wireless sensor networks.

KEYWORDS

- **wireless foundations**
- **microelectronics**
- **synchronous hubs**
- **blast memory**
- **bordering centers**
- **sensor systems**

REFERENCES

1. Cohen, R.; Boris, K. Continuous Neighbor Discovery in Asynchronous Sensor Networks. *IEEE/ACM Trans. Network.* **2011,** *19* (1), 69–79.
2. Peda Gopi, A.; Suresh Babu, E.; Naga Raju, C.; Ashok Kumar, S. Designing an Adversarial Model against Reactive and Proactive Routing Protocols in MANETS: A Comparative Performance Study. *Int. J. Electr. Comput. Eng.* **2015,** *5* (5), 1111–1118.
3. Essa, I. A. Ubiquitous Sensing for Smart and Aware Environments. *IEEE Pers. Commun.* **2000,** *7,* 47–49.
4. Guidoni, D. L.; Boukerche, A.; Villas, L. A.; Souza, F. S. H.; Mini, R. A. F.; Loureiro, A. A. F. In *A Framework Based on Small World Features to Design HSNS Topologies with QoS.* IEEE Symposium on Computers and Communications (ISCC), 2012, July 1–4, 2012; IEEE: Cappadocia, Turkey, 2012.
5. Haartsen, J. *Bluetooth Baseband Specification v.1.0.* www.bluetooth.com.
6. Salonidis, T.; Bhagwat, P.; Tassiulas, L.; LaMaire, R. O. In *Distributed Topology Construction of Bluetooth Personal Area Networks.* Proceedings IEEE INFOCOM, 2001; pp 1577–1586.
7. You, L.; Yuan, Z.; Yang, P.; Chen, G. *ALOHA-like Neighbor Discovery in Low-duty-cycle Wireless Sensor Networks 978-1-61284-254-7/11*; IEEE, 2011.
8. Khanmirza, H.; Landsiedely, O.; Papatriantafilouy, M.; Yazdani, N. *Evaluating Passive Neighborhood Discovery for Low Power Listening MAC Protocols*; IEEE, 2014.
9. Iyer, V.; Pruteanu, A.; Dulman, S. *NetDetect: Neighborhood Discovery in Wireless Networks Using Adaptive Beacons 978-0-7695-4542-4/11*; IEEE, 2011. DOI: 10.1109/SASO.2011.14.
10. Kohvakka, M.; Suhonen, J.; Kuorilehto, M.; Kaseva, V.; Hännikäinen, M.; Hämäläinen, T. D. Energy-efficient Neighbor Discovery Protocol for Mobile Wireless Sensor Networks. *Ad. Hoc. Netw.* **2009,** *7* (1), 24–41.
11. Chen, L.; Fan, R.; Bian, K.; Chen, L.; Gerla, M.; Wang, T.; Li, X. In *On Heterogeneous Neighbor Discovery in Wireless Sensor Networks 2015*, IEEE Conference on Computer Communications (INFOCOM), 2015; pp 693–701. DOI: 10.1109/INFOCOM. 2015.7218438.
12. Sun, G.; Wu, F.; Chen, G. *Neighbor Discovery in Low-duty-cycle Wireless Sensor Networks with Multipacket Reception*; IEEE, 2012. DOI: 10.1109/ICPADS.2012.73.
13. Huang, T.; Chen, H.; Zhang, Y.; Cui, L. EasiND: Effective Neighbor Discovery Algorithms for Asynchronous and Asymmetric-duty-cycle Multi-channel Mobile WSNs. *Wireless Pers. Commun.* **2015,** *84,* 3031–3055. DOI: 10.1007/s11277-015-2781-8.

CHAPTER 12

HYBRID OVERLAY/UNDERLAY TRANSMISSION: AN EFFICIENT MECHANISM TO ENCOURAGE PRIMARY USERS TO COOPERATE WITH SECONDARY USERS

C. S. PREETHAM[1*], M. SIVAGANGA PRASAD[2], and T. V. RAMAKRISHNA[1]

[1]Department of Electronics and Communication Engineering, KL University, Guntur, Andhra Pradesh, India

[2]Department of Electronics and Communication, KKR and KSR Institute of Technology and Sciences, Guntur, Andhra Pradesh, India

*Corresponding author. E-mail: Sunil_veena10@yahoo.co.in

ABSTRACT

One of the fundamental assumptions while deploying a cognitive radio network is that the primary users are willing to tolerate additional interference generated by secondary users and are willing to share their spectrum resource with secondary users. So, the general question that arises here is that why the primary users should tolerate additional interference and share resources? In this chapter, we have presented a new scheme that can provide additional SNR for the primary users by allowing secondary users to coexist. Hybrid overlay/underlay strategy is the proposed scheme in which the idle secondary users are used to maximize the throughput of the primary users. The features of both underlay and overlay are incorporated into this hybrid transmission technique such that the both the licensed and unlicensed users get benefited, as their communication link is uninterrupted. Also, we presented a resource allocation algorithm for the hybrid overlay/underlay transmission method and the performance of the opportunistic and partial relay selection is investigated. This resource allocation algorithm is used for selection of the best relay, power allocated to the relay, and best channel from relay to destination. Simulation results are presented to demonstrate the performance of the proposed hybrid technique over the traditional methods. The effectiveness of usage of the partial and opportunistic relay selection techniques is compared and the conclusions are drawn.

12.1   INTRODUCTION

With the widespread wireless communication technologies and demand for excessive spectrum, the need for intelligent wireless system that changes their mode of operation by being aware of its surrounding environment, for better spectrum utilization, is inevitable. The need for such a system is because of the inefficiency of the traditional spectrum management to utilize the spectrum efficiently. In the traditional spectrum management, the spectrum is allocated for various wireless standards or licensed users for their exclusive use. One of the technologies that have caught the attention of many researchers is cooperative spectrum-sharing systems. Cognitive radio (CR) and cooperative diversity (CD) are the two techniques that are used in this technology. Efficient spectrum utilization is provided by CR and CD improves the reliability of communication. Usually in such cooperative communication systems, there are several nodes in between the source and destination to be chosen as the relay node. The performance of such system can be improved by selecting one of the nodes as the best relay, based on a relay selection criterion. The classification of the relay selection schemes can be done into two types, partial relay selection and opportunistic relay selection. In the partial relay selection, a relay is chosen as the best relay based on the signal-to-noise ratio (SNR) at the first hop or at the second hop. The relay that has highest SNR at any one of the hop is chosen as the best relay. In the opportunistic relay selection, the relay is selected based on the end-to-end SNR value. The relay with highest end-to-end SNR value is chosen as the best relay.

To utilize the free spectrum bands that are licensed to the primary users (PUs), the secondary users (SU) employ two spectrum access strategies, underlay and overlay strategies. In underlay CR strategy, a tolerable interference level is set at the PU. The SU is allowed to use the spectrum as long as the interference generated by the SU is below this predetermined interference threshold (interference temperature). In overlay CR strategy, the SU node is used to relay the data between two PUs. The SU node can use majority of its power to transmit PU data and the rest of power to transmit its own data. The interference generated at the PU due to the transmission of SU data is compensated by the additional SNR offered by using SU node as relay. The combined access strategy in CR has been a topic of interest recently. This is done to exploit the advantages of both overlay and underlay access strategies. These combined access strategies are predominantly called as hybrid CR systems. A hybrid underlay/overlay transmission scheme with the aim of achieving better statistical delay quality of service provisioning is presented in Ref. [1]. The CR is designed to switch between overlay and underlay based on the operating condition of PU in Refs. [2–5]. In Ref. [6], the authors have presented a hybrid CR scheme with energy-harvesting ability. A distributed power allocation algorithm in which the multiple SUs use a common node as relay and compete for the power of relay to transmit their signals is presented in Ref. [7]. In Ref. [7], the SUs transmit with different spectrum-sharing modes. The opportunistic relay selection performance has been analyzed for the case of dual hop transmission in Refs. [8, 9]. Performance of partial relay selection in dual hop communication systems with semiblind relaying is studied in Ref. [10]. The end-to-end performance of the cooperative AF relaying by using opportunistic relay selection and partial relay selection has been detailed and selection criteria have been presented in Ref. [11].

Unlike the previous hybrid CR algorithms, instead of switching between the overlay and underlay transmission modes, we consider combining the feature of both the modes and propose a novel transmission technique. This hybrid overlay/underlay method beneficiates both the PUs and SUs. This technique uses the idle SUs as the relay node for the transmission of the PU signals. The SUs that have data to send will transmit in underlay mode without causing any interference to PUs. We consider the selection of best node, interference from the active SUs, transmission power of the best relay selected to improve the throughput of the PU. The PUs are benefited with the additional SNR provided by utilizing the idle relays and the SUs are benefited as they are able to utilize the PU band in underlay mode. Also, analysis on selection of the best relay using opportunistic relay selection and partial relay selection is presented.

The remainder of this paper is organized as follows. In Section 12.2, we introduce the CR system with the PUs and SUs in which the opportunistic and partial relay selection must be implemented. In Section 12.3, we investigate the performance of the hybrid CR scheme and relay selection techniques. Specifically, the analysis is done on the selection of best relay using partial or opportunistic relay selection, power allocated to the best relay, and efficient channel from relay to destination. Section 12.4 presents the simulation environment and the performance comparisons of the selection techniques. Finally, the concluding remarks are drawn in Section 12.5.

12.2 SYSTEM MODEL

Consider a CR network that comprises one primary transmitter (source), one primary receiver (destination), "L" idle SUs, and "M" active SUs. Initially, the source has direct link with destination. The relayed path is chosen when one of the relays among the "L" idle SUs can give higher SNR than the direct path. This model is depicted in Figure 12.1. The proposed technique can identify the best relay amongst the idle SUs; the active SUs operate in usual underlay mode. This operation of the active SUs in underlay mode causes additional interference to idle SUs and primary receiver. This is shown by a dotted line in Figure 12.1. There are "K" channels available from SU relays to the PU destination.

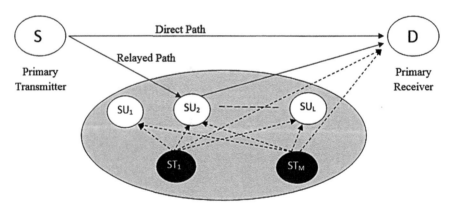

FIGURE 12.1 System model.

In the proposed hybrid overlay/underlay transmission mode, the active SUs operate in underlay transmission mode; the selected relay node also works in underlay mode so that the other PU transmissions in the vicinity don't get disturbed. The PU sender and receiver will use the idle SUs as the relay to transmit their signals. This idle SU will use its entire power for PU transmission. In the ordinary overlay transmission, the SU uses only part of its power for PU transmission. So in the proposed hybrid overlay/underlay transmission scheme, we have the advantageous features of both overlay and underlay. Here, both the PUs and SUs can transmit their signals with full potential, since we have utilized the idle SUs for PU transmission.

Reactive relay selection[12] is performed at the PU receiver. The signals from "L" relays over the K channels arrive at the PU destination. The PU destination node selects the best pair of channel and relays based on the SNR over the link. The pair with the higher SNR is chosen as the relay path. Let the SNR over the first hop is γ_{sr} and the SNR over the second hop is γ_{rd}. The selection criterion in the partial relay selection scheme to select the best relay and channel is

$$(J,K) = \arg \max_{J,K}\{\gamma_{RD}\} \qquad (12.1)$$

The end-to-end SNR is given by $\gamma = \gamma_{sr}\gamma_{rd}/(\gamma_{sr} + \gamma_{rd})$. The selection criterion in the opportunistic relay selection scheme to select the best relay and channel is

$$(J,K) = \arg \max_{J,K}\{\gamma\} \qquad (12.2)$$

12.3 HYBRID OVERLAY/UNDERLAY RELAY SELECTION PROTOCOL

The proposed hybrid overlay/underlay relay selection protocol presents the method for selection of the best relay, best channel, and power of the best relay in the hybrid overlay/underlay model given in Figure 12.1. The various parameters such as interference threshold, PU transmitter power, distance between SUs and PUs, and the spectral distances between SU and PU channels are considered in the protocol. Let $\alpha_{PT_x-PR_x}$, $\alpha_{ST_i-PR_x}$, $\alpha_{PT_x-SU_j}$, $\alpha_{SU_j-PR_x}$, and $\alpha_{ST_i-SU_j}$ are the link gains over links $PT_x \rightarrow PR_x$, $ST_i \rightarrow PR_x$, $PT_x \rightarrow SU_j$, $SU_j \rightarrow PR_x$, and $ST_i \rightarrow SU_j$. Based on the signal-to-interference-plus-noise ratio (SNIR) when the direct link between PU transmitter and receiver is present, we first formulate the primary target rate. Let the distance-dependent path loss factor be n. When P_{PT} is the power transmitted by the PU source, the received power P_{PR} at the PU receiver PR_x is given by

$$P_{PR} = \frac{\alpha_{PT_x-PR_x} P_{PT}}{\left(d_{PT_x-PR_x}\right)^n} \qquad (12.3)$$

The distance between the PT_x and PR_x is denoted by $d_{PTx} - PR_x$. If ST_i ($i = 1, 2, \dots M$) are the active SUs and PST_i is their transmitted power, then the interference power strength P' at the PR_x is expressed as

$$P'_i = \frac{\alpha_{ST_i-PR_x} P_{ST_i}}{\left(d_{ST_i-PR_x}\right)^n} \tag{12.4}$$

where $d_{ST_i} - PR_x$ is the distance between secondary transmitter ST_i and primary receiver PR_x. The SNIR over the link $PT_x - PR_x$ at the primary receiver is defined as

$$SNIR_{PT_x-PR_x} = \frac{P_{PR}}{\sum_{i=1}^{M} P'_i + \sigma_p^2} \tag{12.5}$$

where σ_p^2 is the variance of additive white Gaussian noise (AWGN) on primary transmitter to receiver link. The achievable rate R_{target} bits/s/Hz for the links $PT_x - PR_x$ is given by

$$R_{target} = \log_2\left(1 + SNIR_{PT_x-PR_x}\right) \tag{12.6}$$

This is the target rate achieved over the direct link. The rate achieved over the relayed path after selecting the best relay is now evaluated in the following steps. One of the idle SUs Re_j among the "M" idle relays acts as a best relay. The power received at relay SU_j, if P_{P_T} is the PU source transmitted, is given by

$$P_{SU_j} = \frac{\alpha_{PT_x-SU_j} P_{P_T}}{\left(d_{PT_x-SU_j}\right)^n} \tag{12.7}$$

where $d_{PT_x-SU_j}$ is the distance between the primary transmitter and the idle SU. Additional interference is caused at the idle SUs due to the active SUs. The additional interference power p'_{ij} at SU_j due to active SUs is expressed as[13]

$$p'_{ij} = \frac{\alpha_{ST_i-SU_j} P_{ST_i}}{\left(d_{ST_i-SU_j}\right)^n} \tag{12.8}$$

where p'_{ij} is the interference from user i to user j and the distance between the active SU and the idle SU is $d_{ST_i-SU_j}$. The primary transmitter sends the data to the relays on separate channels. The rate with which the data arrives at the idle SUs is given by

$$R_{P,SU_j} = \frac{1}{2}\log\left(1 + \frac{P_{SU_j}}{\sigma_j^2 + \sum_{i=1}^{M} p'_{ij}}\right) \tag{12.9}$$

where σ_j^2 is the variance of AWGN on primary transmitter to idle SU's link. For every relay, SU_j paired to every subcarrier k calculate the power required to get the same rate of source to relay link in relay to destination link.

$$p_{j,k}^{rate} = \frac{\left(2^{\left(2R_{P,SU_j}\right)} - 1\right)\left(\sigma_k^2 + \sum_{i=1}^{M} p_{ij}'\right)\left(d_{SU_j-PR_x}\right)^n}{\alpha_{SU_j-PR_x}} \tag{12.10}$$

where $d_{SU_j-PR_x}$ is the distance between idle SU and the PU receiver and σ_k^2 is the variance of AWGN on idle SU-to-PU receiver link. The maximum power that can be allocated to each relay for all the combinations of (j,k) is calculated using the following expression:

$$p_{j,k}^{max} = \frac{I_{th}}{\Omega_{j,k}} \tag{12.11}$$

where I_{th} is the interference threshold and $\Omega_{j,k}$ is the interference factor of the channel. The interference factor $\Omega_{j,k}$ is given by

$$\Omega_{j,k} = \alpha T_s \int_{d_k-B/2}^{d_k+B/2} \left(\frac{\sin \pi f T_s}{\pi f T_s}\right)^2 df \tag{12.12}$$

where T_s is the sampling time, α is the gain of the channel, B is the bandwidth occupied by the PU channel,[14] and d_k is the distance in frequency between the subcarrier k and the PU channel. The final power allocated to each relay SU_j over the channel k is[15]

$$power_{j,k} = \min\left(p_{j,k}^{max}, p_{j,k}^{rate}\right) \tag{12.13}$$

The power of the signal received at PU destination from the relay is given by

$$power_{j,k}^{Re} = \frac{\alpha_{SU_j-PR_x} Power_{j,k}}{\left(d_{SU_j-PR_x}\right)^n} \tag{12.14}$$

The optimal channel and relay pair can be selected using opportunistic or partial relay selection.

12.3.1 PARTIAL RELAY SELECTION

The optimal channel and relay is selected using expression (12.15). The pair (j,k) that gets the maximum value of this expression is the optimal pair

$$\left(j^{\text{opt}}, k^{\text{opt}}\right) = \arg \max_{J,K} \left\{ \text{power}_{j,k}^{\text{Re}} \right\} \tag{12.15}$$

The rate of the signal that is received by the PU destination from this optimal pair is given by

$$R_{\text{PR}_x} = \frac{1}{2} \log \left(1 + \frac{\text{power}_{j^{\text{opt}}, k^{\text{opt}}}^{\text{Re}}}{\sigma_k^2 + \sum_{i=1}^{M} p_i'} \right) \tag{12.16}$$

12.3.2 OPPORTUNISTIC RELAY SELECTION

In this relay selection criterion, the optimal pair $(j^{\text{opt}}, k^{\text{opt}})$ is selected based on the end-to-end SNR. To perform this, the PU destination must be aware of full channel state information. The optimal pair is given by[11]

$$\left(j^{\text{opt}}, k^{\text{opt}}\right) = \arg \max_{J,K} \left\{ \frac{P_{\text{SU}_j} \text{power}_{j,k}^{\text{Re}}}{\left(P_{\text{SU}_j} \alpha_{\text{PT}_x-\text{SU}_j} \sigma_j^2 + \text{power}_{j,k \alpha_{\text{SU}_j-\text{PR}_x}}^{\text{Re}} \sigma_k^2 + \sum_{i=1}^{M} p_{ij}' \right)} \right\} \tag{12.17}$$

The rate of the received signal at PU destination by using the optimal pair obtained in opportunistic relay selection is given by

$$R_{\text{PR}_x} = \frac{1}{2} \log \left(1 + \frac{P_{\text{SU}_j} \text{power}_{j^{\text{opt}}, k^{\text{opt}}}^{\text{Re}}}{\left(P_{\text{SU}_j} \alpha_{\text{PT}_x-\text{SU}_j} \sigma_j^2 + \text{power}_{j,k \alpha_{\text{SU}_j-\text{PR}_x}}^{\text{Re}} \sigma_k^2 + \sum_{i=1}^{M} p_{ij}' \right)} \right) \tag{12.18}$$

The direct path is neglected and the transmission is switched to relayed path under any one of the following two conditions.

1. If rate over the relayed path (R_{PR_x}) obtained in either partial relay selection or opportunistic relay selection is greater than the rate over the direct path (R_{target}), that is $R_{\text{PR}_x} = R_{\text{target}}$.
2. If the direct link between source and destination is broken due to severe shadowing or fading.

The power that is received from the PU source to PU destination over the direct link is evaluated by using (12.1). The interference at the PU destination due to the active SUs is given by (12.2). The target data rate R_{target} over the direct link is obtained from (12.3). To find the data rate over the relayed channel, we first need to recognize the best relay and channel. The power of the received signal at the idle SUs from the PU source is given by (12.4). The interference by the active SUs to idle SUs is given by (12.5). From these values, we evaluate the data rate of the signal from SU source to idle SUs using (12.6). To achieve maximum throughput, the data rate over the two hops must be equal.[16] The power required by the relay to achieve the same data rate in hop 1 is given by (12.7). The maximum power that can be used by a relay is given by (12.8). The minimum of the two powers obtained in (12.7) and (12.8) is the power allocated to the relay, expressed in (12.9). The power with which the signal reaches PU destination from idle SUs is given by (12.10). The partial relay selection criterion is expressed in (12.11). This gives the optimal pair of relay and channel from SU relay to PU destination. Rate over the relayed path is given by (12.12). The opportunistic relay selection criterion is expressed in (12.13) and the rate over the relayed path selected by opportunistic relay selection is given by (12.14).

12.4   SIMULATION RESULTS

To validate the proposed transmission technique, simulations were carried out and the results are presented in this section. Simulations were carried out on MATLAB. The PU source and destination are located at (4, 40) and (500, 40). The two active SUs are located at (200, 20) and (400, 20). The five idle SUs are located at (100, 40), (200, 40), (250, 40), (300, 40), and (400, 40). The scenario we considered assumes the power of PU source P_{PT} = 10 dB and the power of active secondary transmitter is P_{ST_1} = P_{ST_2} = 10 dB. Link gain α and path loss factor n are taken as $(0.097/d^2)^{1/2}$ and 2. To prevent the loss of generality, we assume $\sigma_p^2 = \sigma_j^2 = \sigma_k^2 = \sigma^2 = 10^{-13}$.

Figure 12.1 presents the simulation model of the system; it presents the five idle SUs, two active SUs, and PU source and destination. We consider the scenario presented in Figure 12.1, where there are five idle SUs, two active SUs, and a PU sender and receiver. The figure helps to identify the distances between the nodes. Figure 12.2 shows the spectrum allocation of the five available channels from SU relay to PU receiver. The figure helps to find the spectral distances between the channels. Bandwidth of the PU channel is 2 MHz and the bandwidth of relay channels is 1 MHz.

The performance comparison over the direct path and relayed path by using the different transmission schemes is shown in Figure 12.3. The capacity is plotted by varying the interference threshold and the power of PU transmitter is fixed at 10 dB. Since there is no effect of varying interference threshold[17] on overlay transmission and direct path, the capacity response of them is flat. The response with partial relay selection is represented with a dotted line and the response of opportunistic relay selection is represented with a straight line. Figure 12.3 shows that the capacity over the relayed path, chosen by using partial relay selection, gives better performance than the path chosen by using opportunistic relay selection. When the interference threshold is less than 3 mW, overlay transmission gives better capacity than the proposed scheme. From Figure 12.3, it is evident that the proposed scheme gives good performance than the traditional overlay and underlay techniques at an interference threshold $I_{th} \geq 3$ mW, with the relayed path selected using partial relay selection.

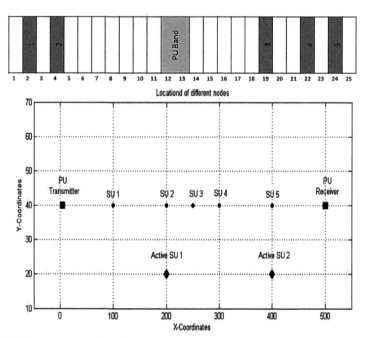

FIGURE 12.2   Spectrum allocation.

The capacity over the direct path and relayed path with varying transmitting power of PU source is studied in Figure 12.4. The performance of the proposed hybrid CR scheme is better than the traditional overlay and underlay at various values of the PU transmitting power. The capacity reaches a saturation value at a power of 15 dB and after this the capacity is constant. It is previously observed in Figure 12.3 that capacity over the relayed path by the use of partial relay selection is better when compared to opportunistic relay selection.

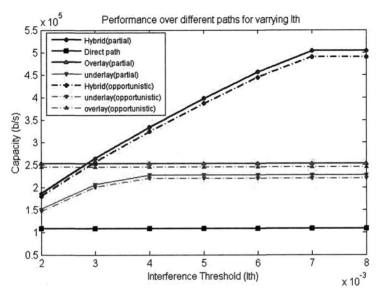

FIGURE 12.3 (See color insert.) Capacity over different paths with partial and opportunistic relay selection and fixed PU transmitting power.

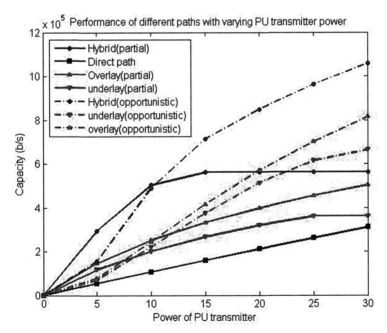

FIGURE 12.4 (See color insert.) Capacity over different paths with partial and opportunistic relay selection and fixed interference threshold.

Figure 12.3 is plotted with the PU transmitter power up to 10 dB. In Figure 12.4, it can be observed that above 10 dB of PU transmitting power, the capacity achieved by opportunistic relay selection is higher than the partial relay selection. It is evident from Figure 12.4 that the opportunistic relay selection achieves significantly higher performance at higher SNR levels.

12.5 CONCLUSION

A novel hybrid overlay/underlay strategy for PU throughput maximization is presented. We have analyzed two different relay selection schemes. An algorithm to allocate various resources in such hybrid CR scheme is presented. We have shown from the simulations that the proposed hybrid overlay/underlay technique has better performance over the traditional overlay and underlay techniques. The throughput achieved by the use of the proposed technique is higher than the throughput achieved on the direct path.

This is a very useful incentive for the licensed users to cooperate with the unlicensed users. We have also presented the performance of the two relay selection criteria. The partial relay selection is preferable up to the PU transmission of 10 dB. Beyond 10 dB, opportunistic relay selection gives best performance. The opportunistic relay selection has good performance at high SNR, but it is complex, as it needs the channel state info of the first hop. Partial relay selection is more preferable, as it is simple and the usual transmission power level of PU is commonly 10 dB. At the PU transmission power of 10 dB, the proposed hybrid overlay/underlay transmission scheme gives optimal performance when combined with partial relay selection.

KEYWORDS

- **wireless communication technologies**
- **traditional spectrum management**
- **partial relay selection**
- **primary users**
- **secondary users**
- **cognitive radio**
- **cooperative diversity**

REFERENCES

1. Soysa, M.; Suraweera, H. A.; Tellambura, C.; Garg, H. K. Partial and Opportunistic Relay Selection with Outdated Channel Estimates. *IEEE Trans. Commun.* **2012,** *60,* 840–850.
2. Tsiftsis, T. A.; Karagiannidis, G. K.; Mathiopoulos, P. T.; Kotsopoulos, S. A. Nonregenerative Dual-Hop Cooperative Links with Selection Diversity. *EURASIP J. Wireless Commun. Network.* **2006,** *2006,* 1–8.
3. Torabi, M.; Ajib, W.; Haccoun, D. In *Performance Analysis of Amplify-and-Forward Cooperative Networks with Relay Selection over Rayleigh Fading Channels.* IEEE 69th, Vehicular Technology Conference, 2009 (VTC Spring 2009), 2009; pp 1–5.
4. Ikki, S.; Ahmed, M. Performance Analysis of Cooperative Diversity Wireless Networks over Nakagami-m Fading Channel. *IEEE Commun. Lett.* **2007,** *11,* 334–336.
5. Suraweera, H. A.; Michalopoulos, D. S.; Karagiannidis, G. K. Semi-Blind Amplify-and-Forward with Partial Relay Selection. *Electron. Lett.* **2009,** *45,* 317.
6. Krikidis, I.; Thompson, J.; McLaughlin, S.; Goertz, N. Amplify-and-Forward with Partial Relay Selection. *IEEE Commun. Lett.* **2008,** *12,* 235–237.
7. Barua, B.; Ngo, H.; Shin, H. On the SEP of Cooperative Diversity with Opportunistic Relaying. *IEEE Commun. Lett.* **2008,** *12,* 727–729.
8. Bletsas, A.; Shin, H.; Win, M. Z. Outage Optimality of Opportunistic Amplify-and-Forward Relaying. *IEEE Commun. Lett.* **2007,** *11,* 261–263.
9. Maham, B.; Hjorungnes, A. Performance Analysis of Amplify-and-Forward Opportunistic Relaying in Rician Fading. *IEEE Signal Process. Lett.* **2009,** *16,* 643–646.
10. da Costa, D. B.; Aissa, S. End-to-end Performance of Dual-Hop Semi-Blind Relaying Systems with Partial Relay Selection. *IEEE Trans. Wireless Commun.* **2009,** *8,* 4306–4315.
11. Minghua, X.; Aissa, S. Cooperative AF Relaying in Spectrum-Sharing Systems: Performance Analysis under Average Interference Power Constraints and Nakagami-m Fading. *IEEE Trans. Commun.* **2012,** *60,* 1523–1533.
12. Bletsas, A.; Shin, H.; Win, M. Cooperative Communications with Outage-Optimal Opportunistic Relaying. *IEEE Trans. Wireless Commun.* **2007,** *6,* 3450–3460.
13. Prasad, M. S. G.; Siddaiah, P.; Reddy, L. P. In *Analysis of Different Direction of Arrival (DOA) Estimation Techniques Using Smart Antenna in Wireless Communications.* Proceedings of the International Conference on Advances in Computing, Communication and Control (ICAC3 '09), 2009; p 639.
14. Bansal, G.; Hossain, M.; Bhargava, V. Optimal and Suboptimal Power Allocation Schemes for OFDM-Based Cognitive Radio Systems. *IEEE Trans. Wireless Commun.* **2008,** *7,* 4710–4718.
15. Preetham, C. S.; Prasad, M. S. G. In *Relay, Power and Subchannel Allocations for Underlay Non-LOS OFDM-Based Cognitive Networks under Interference Temperature.* International Conference on Signal Processing and Communication Engineering Systems (SPACES), Guntur, 2015; pp 205–209.
16. Shaat, M.; Bader, F. Asymptotically Optimal Resource Allocation in OFDM-Based Cognitive Networks with Multiple Relays. *IEEE Trans. Wireless Commun.* **2012,** *11,* 892–897.
17. Manimegalai, C. T.; Kalimuthu, K.; Gauni, S.; Kumar, R. Enhanced Power Control Algorithm in Cognitive Radio for Multimedia Communication. *Indian J. Sci. Technol.* **2016,** *9.*

CHAPTER 13

OFDM-BASED PACKET TRANSCEIVER ON USRP USING LABVIEW

EDURU HEMANTH KUMAR and V. V. MANI*

Department of Electronics and Communication Engineering, National Institute of Technology, Warangal, Telangana, India

Corresponding author. E-mail: vvmani@nitw.ac.in

ABSTRACT

Orthogonal frequency division multiplexing (OFDM) is one of the most used multicarrier technique in today's wireless standards, due to its attractive properties providing robustness to the overall system performance. Software-defined radio (SDR) is becoming a very popular choice for developing, prototyping, and testing various wireless communication architectures. In this chapter, we present the implementation of OFDM-based packet transceiver architecture using universal software radio peripheral (USRP); an SDR platform in LabVIEW. The OFDM symbols are generated in accordance with the IEEE 802.11 specifications with symbols encapsulated in packet format before transmission. The system performance is measured in terms of packet received ratio (PRR) by considering various real-time channel conditions. Results are seemed to be more practical in nature concluding the PRR performance at different gains of the antenna under line of sight (LOS) and non-line of sight (N-LOS) conditions.

13.1 INTRODUCTION

Orthogonal frequency division multiplexing (OFDM) is a versatile multi-carrier technique, dividing the whole band into a number of narrow-band subchannels. Due to its orthogonal nature of the carriers, it adds additional advantage to the system model improving the performance of the system, combating the multipath fading effects in frequency-selective channels. It is the most used modulation technique in a wide range of recent standards such as digital audio broadcasting, digital video broadcasting, WiFi, and long-term evolution. Software-defined radio (SDR) is defined as "radio in which some or all of the physical layer functions are software defined".[1] This gives the user the flexibility of redefining the functioning of the SDR to the application, which is not possible with traditional hardware-based radios (unless through physical modification, thus increasing the cost). This flexibility and cost-effectiveness have motivated the users to shift to SDR for prototyping and testing of their wireless architectures/platforms. The test bed used in this chapter is a universal software radio peripheral (USRP) one of the specific type of SDR, originally developed by Etus Research. USRP developed by National Instruments, the parent company of Etus Research, is becoming the present de facto SDR platform to implement the present transceiver system.

The development of the OFDM architecture that is suitable for implementation on USRP2 boards was investigated in Ref. [2]. In brief, this chapter focuses on the Gnu's Not Unix (GNU) radio implementation of the physical layer of long-term evolution-advanced with simple subcarrier blinding algorithm which is used to reduce the block error rate. The MATLAB implementation of secondary user cognitive link on SDR/USRP was proposed in Ref. [3] with more generic algorithms to improve the overall performance of the OFDM system. More specifically, the performance is studied in terms of channel estimation, time synchronization, and frequency offset compensation. Ref. [4] introduced more generic OFDM system, in which authors quantified the quality of service in terms of packet-received ratio (PRR) using GNU radio for USRP link and C++ for processing the data. Frequency offset errors using Simulink/MATLAB for the signal processing is investigated in Ref. [5]. A LabVIEW-based implementation of basic OFDM system using USRP2 as the test bed is investigated in Ref. [6], evaluating the performance in indoor wireless channels. In this chapter, the focus is on developing an architecture of OFDM and testing the same on a USRP platform. In today's data-centric world, most of the data transmitted is in terms of the packets; thus, PRR is the best quantifier to evaluate the performance of the data communications system, so this chapter has taken PRR as the performance measure. This chapter is structured as follows. Section 13.2 gives an overview of the SDR/USRP followed by Section 13.3 giving the implementation details. These will be followed by the results in Section 13.4 and conclusions and future work in Section 13.5.

13.2   SDR–USRP

USRP is one specific type of SDR in which the baseband operations are configured to run on the host computer, while the front end and high-speed operations such as up-and-down conversions are done in the SDR hardware. USRP-RIO can be considered as an advanced version of the USRP, where RIO has an extra configurable (programmable) field-programmable gate array (FPGA) module; so some (or all) of the time critical/computationally intensive baseband operations can be routed to the FPGA, which differs from the basic USRP device. The basic processing blocks of a USRP are as follows: At the transmitter side, USRP will be interfaced with help of the Ethernet (USRP-RIO was interfaced using PXI) followed by the transmission control block; then, data will be split into I and Q channels; in each of the channels, the data will be first upconverted with the help of digital up converter. Later, the data will be converted to analog with the help of digital to analog converter and then passed through a low-pass filter, and RF conversion is done with the help of mixer and local oscillator. Finally, the transmitted symbols are passed through a transmit amplifier. At the receiver side, the inverse process will be done, that is, first, the received signal is passed through RF amplifier followed by a mixer to convert RF level to IF. Later, a series section of an analog-to-digital converter, digital down conversion, receiver control block, and an Ethernet interface used to connect to a computer. A simple block diagram of USRP is as shown in Figure 13.1.[7] In this chapter, the implementation is done on USRP-RIO without the help of FPGA. The specifications of the USRP-2922 and USRP-RIO-2953R used are mentioned in Tables 13.1[8] and 13.2.[9]

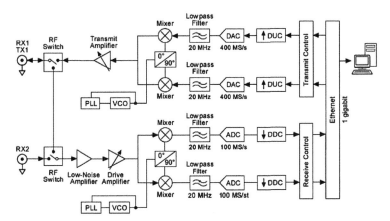

FIGURE 13.1   Block diagram of a USRP.
Source: Reprinted from http://zone.ni.com/reference/en-XX/help/373380B-01/usrphelp/2922_block_diagram/

TABLE 13.1   Specifications of Receiver of NI-USRP-2922.

Parameter	Value
Frequency range	400 MHz–4.4 GHz
Gain range	0–31.5 dB
Frequency accuracy	2.5 ppm
Maximum real-time bandwidth	20 MHz (16 bit sample)
Maximum I/Q sampling rate	25 MS/s (16 bit sample)
DAC	2 ch, 400 MS/s, 16 bit
ADC SFDR	88 dB

Source: Adapted from Ref. [7].

TABLE 13.2   Specifications of NI-USRP-RIO-2953R.

Parameter	Value
Frequency range	1.2–6 GHz
Maximum O/P power (1.2–3.5 GHz)	17–20 dBm
Maximum O/P power (3.5–6 GHz)	7–15 dBm
Gain range	0–31.5 dB
Frequency accuracy	2.5 ppb
Maximum real-time bandwidth	40 MHz
Maximum I/Q sampling rate	200 MS/s
DAC	2 ch, 100 MS/s, 16 bit
ADC SFDR	80 dB

13.3   IMPLEMENTATION

Figures 13.2 and 13.3 show the real-time experimental setup. In the following sections, a detailed explanation of the transmitter and the receiver will be explained.

13.3.1   TRANSMITTER

In this section, the input data can be chosen in two ways, either reading from a user chosen file or generated using a random pseudo noise (PN) sequence. The data are modulated using BPSK/QPSK/8-PSK schemes which can be configured before each run and serial-to-parallel conversion of the data is done, as shown in Figure 13.1. The resultant data symbols are given to IFFT block as IFFT gives the mathematical equivalence of the OFDM symbols mapping over various carriers and IFFT is more mathematically feasible. The OFDM symbol mapping is done in accordance with the IEEE 802.11b specifications; the data are split into 52 of 64 carriers (-26 to -1 and $1–26$); the center carrier was nulled and cyclic prefix was added.

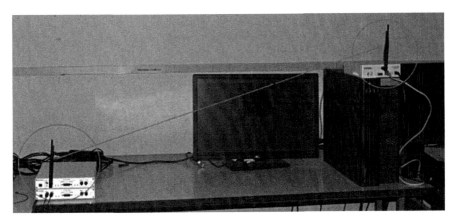

FIGURE 13.2   USRPs separated with line of sight.

FIGURE 13.3   USRPs separated with obstacles in-between them.

The obtained data are to fed to the packet generator block. The packet/frame format is shown in Figure 13.4 in which the OFDM symbol is encapsulated along with guard band bits, sync bits, and packet number, and frame check sequence (FCS) is attached at the end of the packet format where each size is given at each run. The details of the packet/frame are as follows.

Then, frame structure is as shown in Figure 13.4. The sizes of sync and guard bits have to be taken on the trade-off between the coding efficiency and probability of error. The sync bits are generated by using a PN sequence generator order of which has to be specified before the run; this will be useful to find the start of the frame. Guard bits are used to overcome the effects of the overlapping of the successive and previous frames. The packets will be additionally attached with packet sequence so that if they arrive not in order or say a frame incorrectly received, we may be able to retransmit the missing frame (although as of now this feature, i.e., automatic request for repetition, is not implemented). Additionally, a FCS is also added to check the integrity of received frame/packet. The system parameters considered for simulation are shown in Table 13.3 and the frame parameters are shown in Table 13.4.

Guard Band	Sync Data	Packet Number	Data	FCS
0-20	0-20	16	1-2560	16

FIGURE 13.4 Packet format, the numbers below indicate the size of the parameter in bits (not to scale).

TABLE 13.3 Specifications of the OFDM System Implemented.

Parameter	Value
Carrier frequency	1.2–2.4 GHz
Number of subcarriers	64
Null carriers	12
Modulation scheme used	BPSK and QPSK
Cyclic prefix length	16 bits

Source: Adapted from Ref. [7].

TABLE 13.4 Specifications of the Packet.

Parameter	Value (bits)
Guard bits	0–20
Sync bits	0–20
Message size	1–2560
Frame/packet number	16
Frame check sequence	16

Source: Adapted from Ref. [7].

The receiver and transmitter will be running on different machines; there is always a possibility that before the receiver starts its operation, transmitter may have completed the transmission. To avoid this situation, the transmitter is made to run continuously; however in future, it will be modified slightly that once the receiver recovers all the frames, it will send a positive ACK and then the transmitter will stop sending. We may improve this by using other protocols like Go BackN, etc.; then, the frames will be reshaped using the matched pulse shaping filters; there were options to choose among root raised cosine, Gaussian, etc., and their corresponding parameters such as filter length, roll-off factor, etc. Thus, generated frames are fed to the USRP with the help of USRP hardware driver (UHD). There are options to configure the IQ rate, central frequency of the transmission, gain of the antenna, and port of the antenna which are to be transferred.

13.3.2 RECEIVER

The data will be received by USRP and buffered locally. The parameters of the frame and the UHD have to be same as that of the transmitter. The channel estimation and equalization are done and then received data are checked for the sync bits so as to identify the start of the frame. Then, the frames will be ripped of the guard and sync bits and integrity of the frame will be checked with the help of FCS; if integrity fails, the frame will be dropped. If the frame was correct, then the frame number will be stored in those received correctly (this is to avoid any retransmitted packets processing and to check all the packets recovered correctly or not). After all the packets/frames are recovered, the data portion is extracted and the data are sent to fast fourier transform (FFT) modulator and symbols will be mapped to the corresponding data after estimation (minimum least square estimate). Then, data will be regrouped from the block sand displayed and can be written to a file. Additionally, the constellation of the received pattern and spectrum of the data received are calculated. The block diagram was as shown in Figure 13.5.

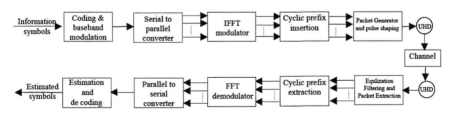

FIGURE 13.5   Architecture of the OFDM implementation.

The experimental setup and procedure: A USRP (NI USRP 2922[8]) is used as receiver and another USRP (NI USRP RIO-2953R[9]) is used as transmitter; they are placed in various positions such as the following:

- Both USRPs have a line-of-sight communication (a very optimistic situation); Figure 13.2.
- Both USRPs do not have any line-of-sight communication and are separated by typical office equipment like PCs, tables, files, and a wall (a more typical situation, which is often the case of a typical office WiFi); Figure 13.3.

The data were transmitted in both the situations with two different frequencies, namely, 2.45 GHz (same as that of WiFi) and 1200 MHz. In each of these situations, the gain of the antennas, modulation schemes, and filter parameters varied.

13.4   RESULTS

The main quantifying factor that to be used in this chapter is PRR; it was shown in Figure 13.6. PRR was plotted against the gain of the receiving antenna. The plot is obtained as follows: the experiment was repeated about 50 times, at each gain level. However during this processes, the receiver has been restricted to operate only for a fixed amount of duration (if it was allowed to operate with anytime restriction, then the receiver will stop only when all the packets are received). Every time, the PRR was observed and the data were averaged and plotted with the help of MATLAB. Throughout this process, all of our electronic equipment were allowed to operate as usual; some of them are connected to WiFi which is operated at 2.4 GHz. It may be observed from the plot (Fig. 13.6) that with the increase in the gain of the antenna, the PRR has been increased, but it may also be seen that there are few times (e.g., 8–9 dB) even though the gain is increased, PRR has been decreased. This is due to a sudden drop of packets for some iterations (thus decreasing the average PRR); it was shown in Figure 13.7 for the case of 9 dB gain and transmitter and receiver separated with obstacles and no line of sight. This sudden drop may be attributed to the remaining electronic equipment which may have actively being operated at the same frequency level. The results seem to more inlay with the predicted values after removing the data which were not within the 5% deviation from the average. The plot in Figure 13.6 corresponds to the parameters listed in Table 13.5. Figures 13.8 and 13.9 show the spectrum and constellation of the OFDM signal; they are inlay with the theoretical predictions within the experimental accuracy. It is also worth mentioning that as we have implemented an FCS, the BER after decapsulation is almost zero irrespective of the modulation scheme used (this is because the frame will not be processed if FCS check fails).

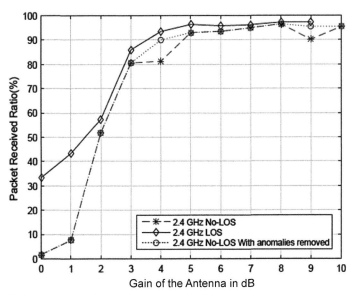

FIGURE 13.6   Packet received ratio versus gain of the antenna.

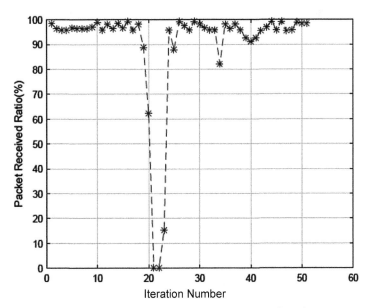

FIGURE 13.7   Variation of the packets received during multiple iterations.

TABLE 13.5   Parameters of the plot in Figure 13.6.

Parameter	Value
Guard bits	16 bits
Distance between Tx and Rx	135 cm
Modulation scheme	QPSK
Sync bits	20 bits
Message size	512 bits
Frame/packet number	16 bits
Frame check sequence	16 bits

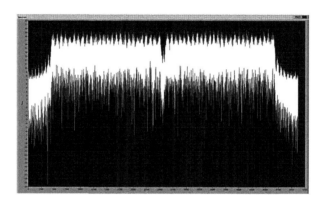

FIGURE 13.8   Spectrum of the OFDM symbols.

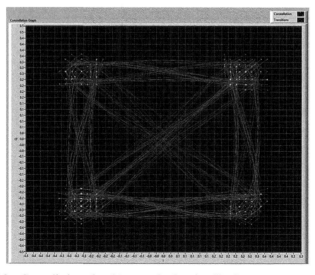

FIGURE 13.9   Constellation after the root raised cosine filtering.

13.5 CONCLUSION AND FUTURE WORK

In this chapter, we have demonstrated a very basic packet-based protocol for transferring the data between two USRPs, but this protocol is very ineffective for practical systems as the retransmission is done without the knowledge of the receiver. This creates unnecessary redundancy of the frames. The basic protocol may be improved by the use of two-way acknowledgment sharing and retransmission of only the packets that are lost or corrupted. In future, we will be implementing these protocols to improve the performance of the system. Although we have repeated the experiment at 1.2-GHz frequency, the results seem to deviate from the expected results; we are yet to investigate them.

KEYWORDS

- **OFDM**
- **SDR**
- **USRP**
- **LabVIEW**
- **packet transceiver**

REFERENCES

1. SDR Forum. *SDR Forum Document Number SDRF-01-S-0006-V2.00* [online]. Available: http://goo.gl/2dttRh.
2. Berardinelli, G.; Zetterberg, P.; Tonelli, O.; Cattoni, A. F.; Srensen, T. B.; Mogensen, P. An SDR Architecture for OFDM Transmission over USRP2 Boards. In *2011 Conference Record of the Forty Fifth Asilomar Conference on Signals, Systems and Computers (ASILOMAR)*, Nov. 2011; pp 965–969.
3. Janji, M.; Brkovi, M.; Eri, M. Development of OFDM Based Secondary Link: Some Experimental Results on USRP N210 Platform. In *Telecommunications Forum (TELFOR)*, 21st Nov. 2013; pp 216–219.
4. Marwanto, A.; Sarijari, M. A.; Fisal, N.; Yusof, S. K. S.; Rashid, R. A. Experimental Study of OFDM Implementation Utilizing GNU Radio and USRP–SDR. In *Communications (MICC), 2009 IEEE 9th Malaysia International Conference on*, Dec. 2009; pp 132–135.
5. Tichy, M.; Ulovec, K. OFDM System Implementation Using a USRP Unit for Testing Purposes. In *Radioelektronika (RADIOELEKTRONIKA), 2012 22nd International Conference*, April 2012; pp 1–4.

6. Lei, L.; Song, C.; Zhang, T. Performance Evaluation for OFDM Link Based on Lab VIEW and USRP. In *Instrumentation and Measurement, Computer, Communication and Control (IMCCC), 2014 Fourth International Conference on*, Sept. 2014; pp 897–901.

7. *NI USRP-2922 Block Diagram* [online]. Available: http://goo.gl/DTfYiK.

8. *NI USRP-2922 National Instruments* [online]. Available: http://sine.ni.com/nips/cds/view/p/lang/en/nid/212997.

9. *USRP-2953R—National Instruments* [online]. Available: http://sine.ni.com/nips/cds/view/p/lang/en/nid/213005.

PART III
Very Large-Scale Integration

CHAPTER 14

AN EFFICIENT SYSTEM DESIGN FOR A 32 BIT SUM-PRODUCT OPERATOR IN MODIFIED BOOTH FORM USING FUSION TECHNIQUE

T. LALITH KUMAR and N. SHEHANAZ[*]

Annamacharya Institute of Technology & Sciences, C.K. Dinne, Kadapa 516003, Andhra Pradesh, India

[*]*Corresponding author. E-mail: shehanazbegum467@gmail.com*

ABSTRACT

The efficient system design for 32-bit addition and then multiplication operation using Radix-2-based modified booth (MB) form by fusion technique is evaluated as explained in this chapter. The MB fusion technique is versatile and genuine technique which is used to generate the reduced partial products for design of larger parallel multipliers. In this work the software format of MB scheme is shown by the schematics in register transfer logic. The simulation and synthesis results are shown by using latest version of Xilinx ISE Design Suite tool and the results are obtained by running Verilog code by checking behavioral model syntax. The implementation of this 32-bit adder and multiplier is shown by using hardware description language.

14.1 INTRODUCTION

Fast adders and multipliers are most important aspects of digital signal processing (DSP) systems. The important part of multiplication is partial product generator, which takes more time to generate the result. The multiplication process was ordinarily applied through a sequence of addition, subtraction, and shift operations. Multiplication will also be considered as continuous procedure of repeated additions. The number to be delivered is the multiplicand, the number of times that it's introduced is the multiplier, and the effect is the product. The generation of partial products is the resultant of step-by-step addition. Generally, the size of multiplicand and multiplier is same in most of the computers.

14.2 EXISTING DESIGN

This system consists of two blocks, first one is an adder block and the second one multiplier block which shows the relation as $Z = X \times (A + B)$. In this existing system, adder result of the add–multiply operator is given as second input and the first input is taken in parallel to the multiplier (Fig. 14.1).[1]

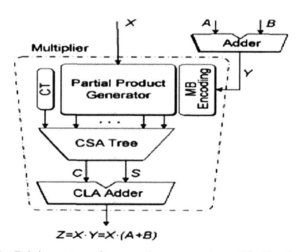

FIGURE 14.1   Existing system of sum-product operator in modified booth form. Adapted from Ref. [1].

14.2.1 MODIFIED BOOTH TECHNIQUE

Modified booth (MB) is an efficient technique form that is used for both signed and unsigned multiplication. In this chapter, radix-4-based truth table[2] is used for multiplying the given two numbers (Fig. 14.2).

The main use of MB form is that it reduces the partial products by half number in multiplication when compared to other radix representation (Table 14.1).[2]

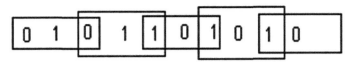

FIGURE 14.2   Grouping of bits in multiplier block using MB form. Adapted from Ref. [2].

TABLE 14.1   Modified Booth Encoding Scheme.

Binary			y_j^{MB}	MB encoding			Input carry
Y_{2j+1}	Y_{2j}	Y_{2j-1}		Sign = s_j	$X1 = 1j$	$X2 = 2j$	$C_{in,j}$
0	0	0	**0**	0	0	0	0
0	0	1	**+1**	0	1	0	0
0	1	0	**+1**	0	1	0	0
0	1	1	**+2**	0	0	1	0
1	0	0	**−2**	1	0	1	1
1	0	1	**−1**	1	1	0	1
1	1	0	**−1**	1	1	0	1
1	1	1	**0**	1	0	0	0

MB, modified booth.

14.2.2   CSA TREE

Carry save addition (CSA) means carry save adder that operates on conditional sum adder concept. Sum and carries are evaluated by considering the input carry as 1 and 0 whatever the carry is generated. When original carry[3] is generated, then the original values of sum and carry are selected by using a multiplexer (Fig. 14.3).[3]

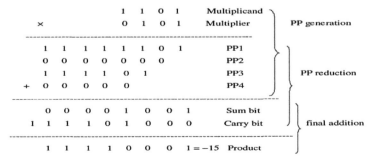

FIGURE 14.3   Example of carry save addition (CSA) tree. Adapted from Ref. [2].

14.2.3   CLA ADDER

A carry-look-ahead adder (CLA)[4] is the fastest adder used in digital circuits (Fig. 14.4).

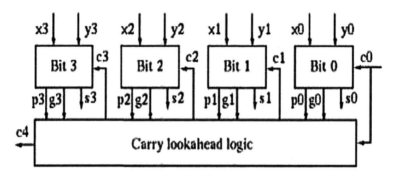

FIGURE 14.4   Block diagram of carry-look-ahead adder. Adapted from Ref. [2].

14.3   PROPOSED DESIGN

The block diagram of proposed design using fusion technique of sum-product operator in MB form using direct recoding concept is ejected in Figure 14.5. In this changed design, the adder circuit is inserted into multiplier circuit by using fusion technique.

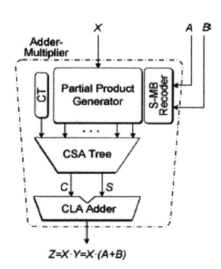

$$Z = X \cdot Y = X \cdot (A+B)$$

FIGURE 14.5   Proposed design of sum-to-modified booth recoder.

14.4 SUM-TO-MODIFIED BOOTH RECODING TECHNIQUE

In this sum-to-modified booth (SMB) scheme, we change the sum of two consecutive digits of input A with two consecutive bits of input B[5] into only one MB digit as Y.

14.4.1 SMB1 RECODING SCHEME

In these schemes, the first one is the important scheme that is shown in Figure 14.6. The implemented first recoding technique is called SMB1 recoding scheme. This technique is explained in detail for both even and odd bit width of input numbers in below explanations.

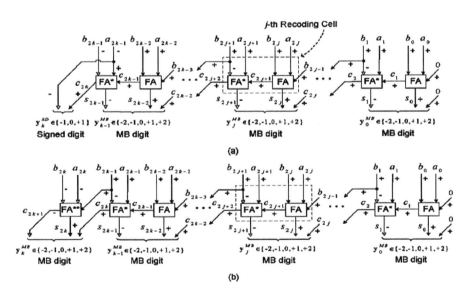

FIGURE 14.6 SMB1 recoding scheme for (a) even and (b) odd number of bits.

14.4.2 SMB2 RECODING SCHEME

The second essential approach of this SMB scheme is referred as SMB2 that is shown in Figure 14.7. SMB2 is evaluated for both even and odd number of inputs. Initially, it considers the values as $c_{0,1} = 0$ and $c_{0,2} = 0$.

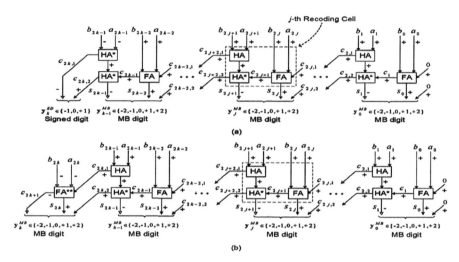

FIGURE 14.7 SMB2 recoding scheme for (a) even and (b) odd number of bits. Adapted from Ref. [4].

14.4.3 SMB3 RECODING SCHEME

The last and final scheme is the combination of both designed half adder (HA) and full adder (FA) which is taken as SMB3.[6] It is explained in detail along with figures for both even and odd number of inputs (Fig. 14.8).[4] At starting, the system considers the carry values as $c_{0,1} = 0$ and $c_{0,2} = 0$.

14.5 PERFORMANCE EVALUATION

14.5.1 THEORETICAL ANALYSIS

The theoretical generation and practical observation of both existing and proposed systems in terms of occupied area and delay in critical path are shown in Figure 14.9 and below tables. This analysis of circuits is based on the unit gate model. More specifically, for the quantitative comparisons, the two-input primitive gates (NAND, AND, NOR, OR) count as one gate equivalent for both area and delay, whereas the two-input XOR, XNOR gates count as two gate equivalents.[7] The area of FA and an HA is 7 and 3 gate equivalents, respectively. The delays of the sum and carry outputs of an FA are 4 and 3 gate equivalents, respectively, while those of an HA are 2 and 1.[8] All the aforementioned information is summarized in Table 14.2.

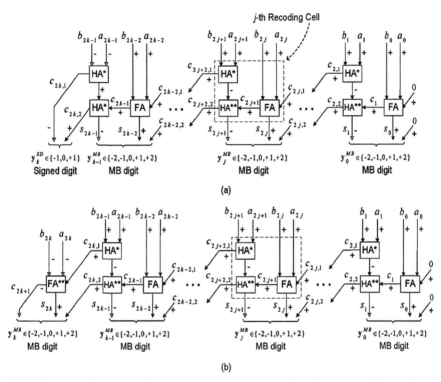

FIGURE 14.8 SMB3 recoding scheme for (a) even and (b) odd number of bits. Adapted from Ref. [4].

TABLE 14.2 Area and Delay of Various Components in the Unit Model.

Components	Area (gate equivalents)	Delay (gate equivalents)
NAND-2, NOR-2	A_g	T_g
NAND-3, NOR-3	$2A_g$	$2T_g$
XOR, XNOR	$2A_g$	$2T_g$
HA	$3A_g$	$T_{HA,carry} = T_g$
		$T_{HA,sum} = 2T_g$
FA	$7A_g$	$T_{FA,carry} = 3T_g$
		$T_{FA,sum} = 4T_g$

Table 14.3 summarizes the area complexity and the critical delay of the proposed recoding schemes SMB1, SMB2, SMB3, and the existing scheme.

TABLE 14.3 Area and Delay of Existing and Proposed Recoding Schemes.

Design	Area complexity	Critical delay
Conventional	$6A_{HA} + 2A_{xor} + A_g = 23A_g$	$T_{HA,sum} + T_{HA,sum} + T_{HA,sum} + T_{xor} + T_g = 9T_g$
SMB1	$2A_{FA} + 2A_{xor} + A_g = 19A_g$	$T_{FA,carry} + 2T_g + T_{xor} + T_g = 8T_g$
SMB2	$A_{FA} + 2A_{HA} + 2A_{xor} + A_g = 18A_g$	$T_{HA,carry} + 2T_g + T_{HA,sum} + T_{xor} + T_g = 8T_g$
SMB3	$A_{FA} + 2A_{HA} + 2A_{xor} + A_g = 18A_g$	$T_{HA,carry} + 2T_g + T_{HA,sum} + T_{xor} + T_g = 8T_g$

14.6 SIMULATION RESULT

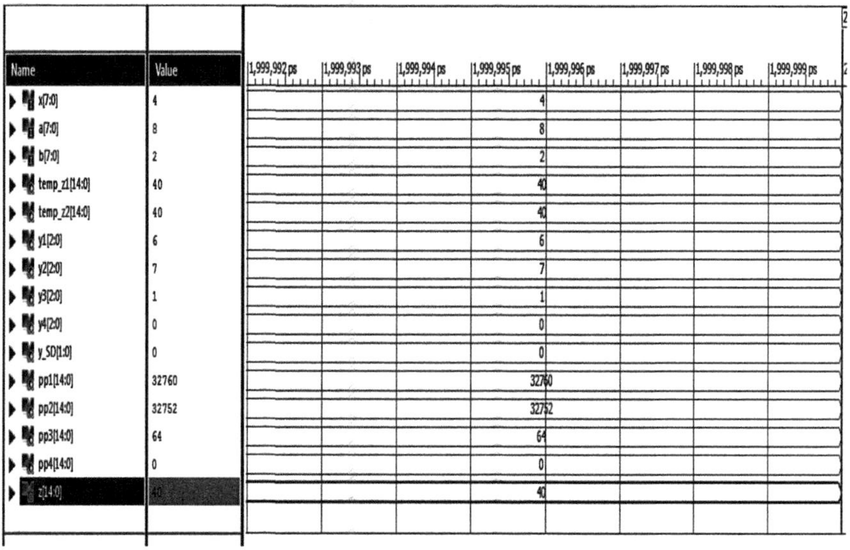

Name	Value	1,999,992 ps	1,999,993 ps	1,999,994 ps	1,999,995 ps	1,999,996 ps	1,999,997 ps	1,999,998 ps	1,999,999 ps
x[7:0]	4				4				
a[7:0]	8				8				
b[7:0]	2				2				
temp_z1[14:0]	40				40				
temp_z2[14:0]	40				40				
y1[2:0]	6				6				
y2[2:0]	7				7				
y3[2:0]	1				1				
y4[2:0]	0				0				
y_SD[1:0]	0				0				
pp1[14:0]	32760				32760				
pp2[14:0]	32752				32752				
pp3[14:0]	64				64				
pp4[14:0]	0				0				
z[14:0]	40				40				

FIGURE 14.9 Simulation result of SMB scheme.

14.7 CONCLUSION

Finally, the authors concluded that in this project, the area occupation in circuit is decreased by fused add–multiply (FAM) operator using fusion technique. Authors designed a structured system for the direct recoding of the sum of two numbers to its MB form. For these supplements, authors discovered the three alternative designs of the proposed SMB constructer and compared them with already existing recoding schemes. When these circuits are designed in FAM operators, they show increased performance criteria when compared with most of the efficient schemes.[9]

The input power consumption may be decreased by the following two conditions that are considered in the future reference for low-power VLSI design.

The bit size may be increased, that is, number of bits considered may be increased in the encoding scheme using MB technique.

The power consumption can be reduced by improving the partial product compression ratio.

KEYWORDS

- **partial product generator**
- **modified booth**
- **carry save addition**
- **adder**
- **block diagram**

REFERENCES

1. Daumas, M.; Matula, D. W. In *A Booth Multiplier Accepting Both a Redundant or a Nonredundant Input with No Additional Delay*. IEEE International Conference on Application-specific Systems, Architectures, and Processors, 2000; pp 205–214.
2. Yeh, W.-C.; Jen, C.-W. High-Speed and Low-Power Split-Radix FFT. *IEEE Trans. Signal. Process.* **2003,** *51* (3), 864–874.
3. Huang, Z.; Ercegovac, M. D. High-Performance Low-Power Left-to-Right Array Multiplier Design. *IEEE Trans. Comput.* **2005,** *54* (3), 272–283.
4. Swartzlander, E. E.; Saleh, H. H. M. FFT Implementation with Fused Floating-Point Operations. *IEEE Trans. Comput.* **2012,** *61* (2), 284–288.
5. Cavanagh, J. J. F. *Digital Computer Arithmetic*; McGraw-Hill: New York, 1984.
6. Xydis, S.; Triantafyllou, I.; Economakos, G.; Pekmestzi, K. Flexible Datapath Synthesis through Arithmetically Optimized Operation Chaining. In *Proc. NASA/ESA Conf. Adaptive Hardware Syst.* 2009; pp 407–414.
7. Bruguera, J. D.; Lang, T. Implementation of the FFT Butterfly with Redundant Arithmetic. *IEEE Trans. Circuits Syst. II, Analog Digit. Signal. Process.* **1996,** *43* (10), 717–723.
8. Available at website: http://www.synopsys.com/Tools/Implementation/SignOff/PrimeTime/Pages/default.aspx [Online].
9. Available at the website: http://www.faraday-tech.com/main/IPonline/category.do?method=showProcessIPList&process=90&categoryID=3089&categoryName=Standard%20Cell [Online].

CHAPTER 15

DESIGN OF AN IMPROVED FAULT COVERAGE PROGRAMMABLE PSEUDORANDOM PATTERN GENERATOR FOR BIST

GADDAM SHRAVAN KUMAR and ADUPA CHAKRADHAR*

Jayamukhi Institute of Technological Sciences, Department of ECE, Warangal 506002, Telangana, India

Corresponding author. E-mail: adupa.chakradhar@gmail.com

ABSTRACT

This project presents a Low power programmable pseudo-random pattern generator which is more suitable for built in self test (BIST) structures used for testing of VLSI circuits. The purpose of the BIST is to reduce power dissipation without affecting the fault coverage. This LP-programmable pseudorandom pattern generator produce pseudo random test patterns with desired toggling level and also enhanced fault coverage compared with other BIST PRPG. It comprised of finite state machine LFSR driving a phase shifter and it allows the device to produce binary sequence with preselected toggling activity. Generator is automatically controlled providing easy and precise tuning. Furthermore, this project introduces a test compression method to avoid repeated pattern generation for testing the same device.

15.1 INTRODUCTION

After a digital circuit has been designed, it is fabricated in the form of silicon chips. The fabrication process is not perfect and due to various reasons, the manufactured circuit in silicon may develop defects that may prevent its correct functioning. A manufacturing test performs the crucial task of identifying those silicon chips that do not function as expected. It involves exercising the functionality of the circuit under test (CUT) by applying appropriate test signals to its inputs and observing the responses.[1] If the responses of the CUT match the expected responses, then the CUT is considered good, else it is labeled as bad. Thus, the goal of testing is to correctly identify a good chip as good and a bad chip as bad. To perform the test, we have to use conventional approach tester. However, a built-in self-test (BIST) technique has been elaborated in which some of the tester functions are incorporated on the chip enabling the chip to test itself.[8] BIST provides a number of well-known advantages. It eliminates the need for expensive testers. It provides the fast location of failed units in a system because the chips can test themselves concurrently; Figure 15.1 is shows the architecture for BIST.

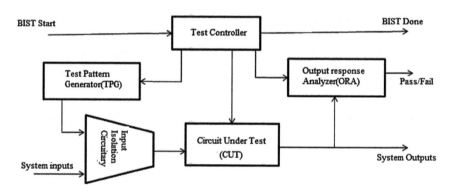

FIGURE 15.1   Architecture of BIST.

To test the circuit, we have to consider that as the CUT. When the test control signal enables, then test pattern generator applies test patterns to the CUT and an output response analyzer (ORA) checks the output responses. ORA will decide whether CUT will pass or fail. When the test control signal disables, testing process is finished.[7] Test pattern generator is the main block in BIST. It generates a set of test patterns to provide high fault coverage to speed up the testing process. The attractive testing approach for BIST is pseudorandom testing. A linear feedback shift register (LFSR) is used to apply pseudorandom patterns to the CUT. An LFSR can also be used as an ORA, thereby serving a dual purpose.

BIST is not sufficient for testing when random patterns are used. To avoid this, we have to modify CUT to make it random pattern testable and also modify the test pattern generator to generate patterns that detect the random pattern resistant faults. This modification enables automated design of pseudorandom BIST implementations that enhance fault coverage and minimize the area.

Pseudorandom pattern generator (PRPG) is the method used as test pattern generator for BIST to enhance fault coverage. An LFSR is used as a PRPG where the input is a linear function of two or more bits. The exclusive-OR (XOR) gate is used in the feedback section of the LFSR to act as PRPG. It contains D flip-flop and XOR gates. The LFSR main part is a shift register. It is used to shift the positions of the content in the register.[3]

This chapter is organized as follows. Section 15.2 introduces existing methods. Preselected toggling (PRESTO) generator and fully operational PRESTO generator with the operation are explained in detail. Section 15.3 introduces proposed work in detail. Section 15.4 gives the simulated results. This chapter concludes with a variety of experimental results and finally wraps up with Section 15.5.

15.2   BASIC ARCHITECTURE OF PRESTO GENERATOR

To test integrated circuits and systems, pseudorandom BIST generators are used. The collection of pseudorandom generators includes LFSRs, cellular automata, and accumulators controlled by a constant value. A large number of random patterns have to be generated for the circuits to hard-to-detect faults. Later, high fault coverage pattern generators are achieved. Generally, we use clock-gating method; two nonoverlapping clocks control the even and odd scan cells of the scan chain to reduce the shift power dissipation.

To reduce the switching activities in scan chain, a pseudorandom BIST scheme was proposed. To detect the faults, we require extra test hardware to store additional deterministic test patterns or by inserting test point into the mission logic. To avoid this problem, an accumulator-based weighted pattern generator technique was introduced. This technique uses one of three weights for test patterns namely 0, 1, and 0.5; therefore, it reduced test application time in accumulator-based test pattern generation.[2]

In this chapter, we propose a PRPG for BIST applications. By using PRESTO levels, we are reducing the switching activity of the generator during scan loading. To reduce the test power, we applying each pattern generated a vector to each PRPG output, which can minimize the input transition, and the number of distinct patterns in a sequence meets the requirement of fault coverage for CUT and the sequence does not contain any repeated patterns. The conventional algorithm of changing the test vectors produced by LFSR contains extra hardware to get more correlated test vectors with a low number of transitions; they reduce the randomness in the patterns but having lower fault coverage and higher test time.[4]

The linear relations are selected from a pattern or consecutive vectors, which is the benefit of using sequential decompressor to generating a sequence. Hence, the proposed test pattern generator (TPG) can be easily implemented by hardware. An n-bit PRPG is connected with a phase shifter feeding scan chain from a kernel of the generator producing the actual pseudorandom test patterns. A PRPG is implemented by using ring generator or LFSR. n-Hold latches are used between the PRPG and phase shifter. An n-bit toggle control register is controlling each stage of the individual latch as shown in Figure 15.2. Latch that enables input is asserted, the given latch is transparent for data going from the PRPG to the phase shifter, and it said to be in toggle mode. When it disables, it captures and saves, for a number of clock cycles, the corresponding bit of PRPG, thus feeding the phase shifter with a constant value. It is now in the hold mode. It is worth noting that each phase shifter output is obtained by XOR-ing outputs of three different hold latches. Therefore, every scan chain remains in a low-power mode provided only disabled hold latches drive the corresponding phase shifter output.

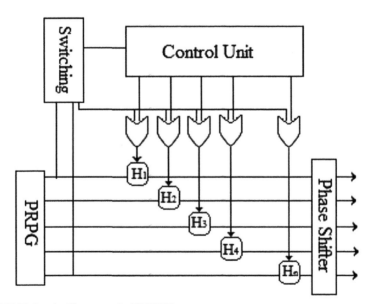

FIGURE 15.2   Architecture of a PRESTO generator.

The contents of toggle registers are zeros (0s) and ones (1s). 1s indicate latches in the toggle mode, thus transparent for data. Their fraction determines a scan switching activity. The control register is reloaded once per pattern with shift register content. The enable signals injected into the shift register are produced in a probabilistic fashion. Using the original PRPG with a programmable set of weights, the weights are determined by four AND gates producing 1s with the probability of 0.5, 0.25, 0.125, and 0.0625, respectively. The OR gate allows choosing of probabilities beyond simple powers of 2. A 4-bit register switching is used to activate AND gates and allows selecting a user-defined level of switching activity. For example, if the switching code is 0100, 25% of the control register will set to 1, and thus 25% of hold latches will be enabled. Given the phase shifter structure, one can assess that the amount of scan chains receives constant values and thus the expected toggling ratio.

An additional 4-input NOR gate detects the switching code 0000, which will switch the LP functionality off. Note that when working in the weighted random mode, the switching level selector ensures statistically the stable content of the control register in terms 1s it carries. As a result, the same fraction of scan chains will stay in the LP mode even when low toggling chains will keep changing from one test pattern to another. It will correspond to a certain level of toggling in the scan chains.

15.2.1   FULLY OPERATIONAL PRESTO GENERATOR

This section presents additional features that make the PRESTO generator fully operational in a wide range of desired switching rate. This approach splits up a shifting period of every test pattern into a sequence of alternating hold and toggle intervals. We use a T-type flip-flop to move the generator back and forth between these two intervals. T flip-flops switch whenever there is a 1 on its input data. If it is set to 0, the generator enters the hold period with all latches temporarily disabled. This is accomplished by placing AND gates on the control register outputs which will allow freezing of all phase shifter inputs. Only a single scan chain crosses a given core. Its abnormal toggling may cause unacceptable heat dissipation that can only be reduced due to temporary hold periods. If the T flip-flop is set to the toggling period, then the latches enabled by the control register can pass data moving from the PRPG to the scan chains. Two additional parameters kept in hold and toggle registers determine how long the entire generator remains either in the hold mode or in the toggle mode. To terminate from modes, a 1 must occur on the T flip-flop input.[3] This weighted pseudorandom signal is produced similarly to that of weighted logic used to feed the shift register. The T flip-flop also controls four 2-input multiplexers routing data from the toggle and hold registers as shown in Figure 15.3. It allows selecting a source of control data that will be used in the next cycle to possibly change the operational mode of the generator.[4]

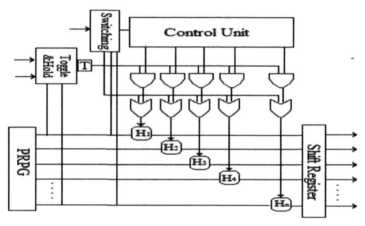

FIGURE 15.3   Fully operational version of a PRESTO generator.

For example, in toggle mode, the input multiplexers observe the toggle register. The flip-flop toggles once the logic output is 1, and as a result, all hold latches freeze in the last recorded state. Until another 1 occurs on the weighted logic output, they will remain in the same state. The occurrence of this event is now related to the content of the hold register, which determines termination of the hold mode.

15.3   IMPROVED FAULT COVERAGE TEST PATTERN GENERATOR

The architecture consists of an additional block transition controller at the output of the phase shifter. The core principle of the decompressor is to disable both weighted logic blocks and deploy deterministic data control. The content of the toggle control register can now be selected in a deterministic manner due to a multiplexer placed in front of the shift register. The toggle and hold registers are employed to alternately reset a 4-bit binary down counter and thus to determine durations of the hold and toggle phases. When this circuit reaches the value of 0, it causes a signal to go high to toggle the T flip-flop. The same signal allows the counter to have the input data kept in the toggle or hold register entered as the next state. Both the down counter and the T flip-flop need to be initialized during every test pattern. The initial value of the T flip-flop decides whether the decompressor will begin to operate either in the toggle or in the hold mode, while the initial value of the counter, further referred to as an offset, determines that mode's duration.[5]

The LP decompressor reduces switching activity during BIST by reducing transitions at scan flip-flops during scan shift operations. Figure 15.4 shows an architecture called LP-decompressor TPG with scan chain to the CUT. The LP-decompressor TPG comprises an r-stage PRPG, a k-input HOLD logics, and a toggle flip-flop (T flip-flop). Hence, it can be implemented with very little hardware.[5]

An immense amount of research has done for BIST test pattern generator to achieve high fault coverage with shorter test application time. Typically, LFSR-based pseudorandom test sequences were modified either by placing a mapping logic between the PRPG outputs and inputs of a CUT or by adjusting the probabilities of outputting 0s and 1s so that the resultant vectors capture characteristics of test patterns for hard-to-detect faults.[6] Test patterns leaving PRPG can also be transformed in a more deterministic fashion, along with the same lines; we will demonstrate that PRESTO-produced LP test patterns are also capable of visibly improving a fault coverage-to-pattern-count ratio. Assuming that the toggle control register can also be driven by deterministic test data (see the location of an additional multiplexer in the front of a shift register in Fig. 15.4), test patterns can be produced with better-than-average fault coverage.[6]

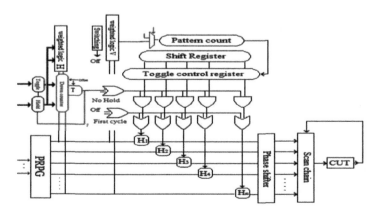

FIGURE 15.4   LP-decompressor TPG with scan chain to the CUT.

15.4   SIMULATION RESULTS

Figure 15.5 shows the simulation of the PRPG. Figure 15.6 shows simulated result of proposed architecture LP-decompressor TPG.

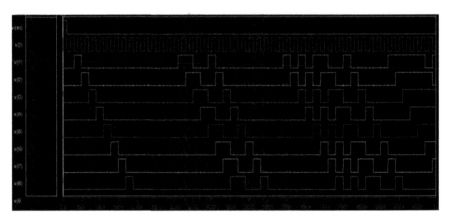

FIGURE 15.5   Pseudorandom pattern generator.

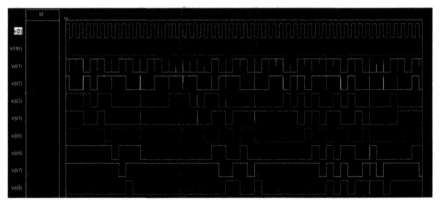

FIGURE 15.6   LP decompressor TPG.

15.5   CONCLUSION

Using PRPG, pseudorandom test patterns are generated with reduced switching activity, and thus, power consumption is reduced. The resultant test vector can yield desired fault coverage faster than conventional pseudo-random patterns, also reducing toggling rates to the desired level. HSPICE is the tool used for simulation. Therefore, an attractive low-power and improved fault coverage test scheme which allows for trading off test coverage, pattern counts, and toggling rates in a very flexible manner. The proposed low power and improved fault coverage PRPG can improve test coverage and reduce pattern count and toggling rates compared to existing TPGs.

KEYWORDS

- **pseudorandom test pattern generators**
- **built-in self-test**
- **switching activity**
- **circuit under test**
- **test data volume compression**

REFERENCES

1. Bushnell, M. L.; Agrawal, V. D. *Essentials of Electronic Testing for Digital, Memory, and Mixed-signal VLSI Circuits.*
2. Filipek, M.; Mukherjee, N. Low Power Programmable PRPG with Test Compression Capabilities. *IEEE Trans. Very Large Scale Integr.* **2015,** *23* (6), 1063–1076.
3. Singh, B.; Khosla, A.; Bindra, S. Power Optimization of Linear Feedback Shift Register (LFSR) for Low Power BIST. In *Proc. IEEE Int. Adv. Comput. Conf. (IACC)*, Mar. 2009; pp 311–314.
4. Krishna, C. V.; Touba, N. A. Reducing Test Data Volume Using LFSR Reseeding with Seed Compression. In *Int. Test Conf.*, 2002.
5. Kavitha, A.; Seetharaman, G.; Prabakar, T. N.; Shrinidhi, S. Design of Low Power TPG Using LP-LFSR. In *2012 Third International Conference on Intelligent Systems Modelling and Simulation*, 2012; pp 334–338.
6. Chetan, J.; Lakkannavar, M. Design of Low Power Test Pattern Generator Using Low Transition LFSR for High Fault Coverage Analysis. In *IJIEEB-V5-N2-3*, 2013; pp 15–21.
7. Saranyadevi, S.; Thangavel, M. A Low Power Structure Design of 2D-LFSR and Encoding Technique for BIST. *Int. J. Adv. Sci. Technol.* **2010,** *18*; 11–22.
8. Santos, M. B.; Braga, J.; Coimbrao, P. *RTL Guided Random-pattern-resistant Fault Detection and Low Energy BIST*. INESC.

CHAPTER 16

DESIGN OF 4-2 COMPRESSOR USING XOR–XNOR BLOCKS FOR HIGH-SPEED ARITHMETIC CIRCUITS

BHARATHA SATEESH[1*] and PRABHU G. BENAKOP[2]

[1]Department of ECE, Vaagdevi College of Engineering, JNTU, Warangal, Telangana, India

[2]Department of ECE, Indur Institute of Engineering and Technology, Ponnala (V), Siddipet, Medak, Telangana, India

*Corresponding author. E-mail: basateesh27@gmail.com

ABSTRACT

A low-power high-speed 4-2 compressor circuit is proposed for fast digital arithmetic integrated circuits. The 4-2 compressor has been widely employed for multiplier realizations to base on a new exclusive OR(XOR) and exclusive NOR(XNOR) module. The proposed circuit shows that power consumption is very less. Power consumption and delay of proposed 4-2 compressor circuit have been compared with earlier reported circuits and proposed circuit is proven to have the minimum power consumption and the lowest delay. Simulations have been performed by using Verilog HDL.

16.1 INTRODUCTION

Multipliers are one of the most significant blocks in computer arithmetic and are generally used in different digital signal processors. There is growing demands for high-speed multipliers in different applications of computing systems, such as computer graphics, scientific calcula-tion, image processing, etc. Speed of multiplier determines how fast the processors will run and designers are now more focused on high speed with low-power consumption. The multiplier architecture consists of a partial product generation stage, partial product reduction stage, and the final addition stage. The partial product reduction stage is responsible for a significant portion of the total multiplication delay, power, and area. Therefore to accumulate partial products, compressors usually implement this stage because they contribute to the reduction of the partial products and also contribute to reduce the critical path which is important to maintain the circuit's performance (Fig. 16.1).[2]

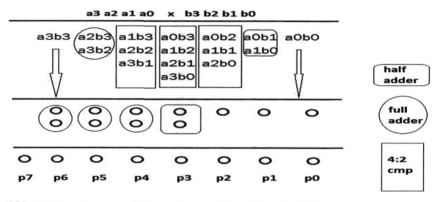

FIGURE 16.1   Structure of 4-2 compressor. Adapted from Ref. [1].

First, all the 8 inputs are fed as input to the AND gates which form 16 products as shown in the figure and form a tree-like structure. Then, these inputs are further fed to half adders,[1] full adders (FAs), and compressors to reduce the partial products. This is accomplished by the use of 3-2, 4-2 compressor structures. A 3-2 compressor circuit is also known as FA cell. Since these compressors are used repeatedly in larger systems, improved design will contribute a lot toward several system performances. The internal structure of compressors is basically composed of XOR–XNOR gates and multiplexers.

The XOR–XNOR circuits are also building blocks in various circuits like arithmetic circuits, multipliers, compressors, parity checkers, etc.[1] Optimized design of these XOR–XNOR gates can improve the performance of multiplier circuit. In this chapter, a XOR–XNOR module has been proposed and 4-2 compressor has been implemented using this module. Using partial product accumulation in proposed circuit reduces power consumption. Following circuit shows that compressor circuit is formed by XOR–XNOR gates.[2]

16.2 COMPRESSOR

One of the major speed enhancement techniques used in modern digital circuits is the ability to add numbers with minimal carry propagation. The basic idea is that three numbers can be reduced to two, in a 3:2 compressor, by doing the addition while keeping the carries and the sum separate. This means that all of the columns can be added in parallel without relying on the result of the previous column, creating a two-output "adder" with a time delay that is independent of the size of its inputs. The sum and carry can be recombined in a normal addition to form the correct result. This process may seem more complicated and pointless, but the power of this technique is that any amount, number of additions can be added together in this manner.[3] It is only the final recombination of the final carry and sum that requires a carry propagating addition. 3:2 Compressor is also known as FA. It adds 3-1 bit binary numbers, a sum, and a carry. The FA is usually a component in a cascade of adders. The carry input for the FA circuit is from the carry output from the cascade circuit. Carry output from FA is fed to another FA.

16.3 4-2 COMPRESSOR

The characteristics of the 4-2 compressor are as follows:

- The outputs represent the sum of the five inputs, so it is really a 5-bit adder as shown in Figure 16.2.
- Both carries are of equal weighting (i.e., add "1" to the next column).
- To avoid carry propagation, the value of C_{out} depends only on A, B, C, and D. It is independent of C_{in}.
- The C_{out} signal forms the input to the C_{in} of a 4-2 of the next column.[8]

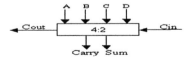

FIGURE 16.2   High level view of the 4-2 compressors.

The common implementation of a 4-2 compressor is accomplished by utilizing two FAs to add binary number cells. 4-2 Compressor is composed of two serially connected FAs. With minimal carry propagation, we use compressor adder instead of other adder. Compressor is a digital modern circuit which is used for high speed with minimum gates which require designing technique.[4] This compressor becomes the essential tool for fast multiplication adding technique on fast processor and lesser area (Fig. 16.3).[3]

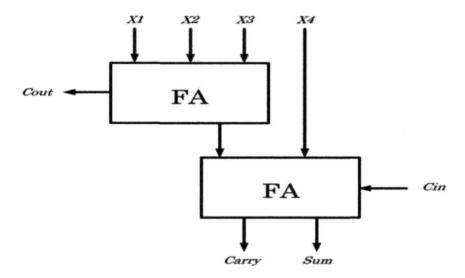

FIGURE 16.3   Structure of 3 bits.

4-2 Compressors are capable of adding 4 bits and one carry, in turn producing a 3-bit output. The 4-2 compressor has four inputs X1, X2, X3, and X4 and two outputs sum and carry along with a carry in (C_{in}) and a carry out (C_{out}). The input C_{in} is the output from the previous lower significant compressor. The C_{out} is the output to the compressor in the next significant stage. The critical path is smaller in comparison with an equivalent circuit to add 5 bits using FAs. However, like in the case of 3-2 compressor, the fact that both the output and its complement are available at every stage is neglected.[5] Thus, replacing some XOR blocks with multiplexers results in a significant improvement[4] with delay as shown in Figure 16.4. Also, the MUX block at the SUM output gets the select bit before the inputs arrive and this minimizes the delay[9] to a considerable extent.

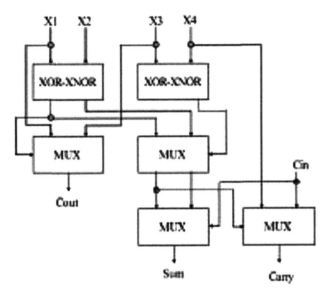

FIGURE 16.4   4-2 Compressor using XOR–XNOR. Adapted from Ref. [4].

16.4   VHDL SIMULATION RESULTS

The proposed 4-2 compressor using XOR–XNOR gates and MUX has been simulated using Xilinx and results are shown in Figures 16.5–16.8[7].

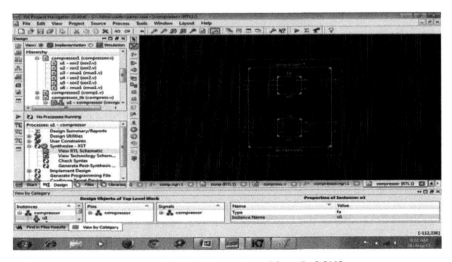

FIGURE 16.5   Compressor using full adders. Adapted from Ref. [10].

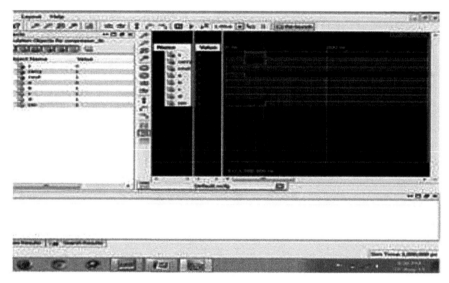

FIGURE 16.6   Schematic of 4-2 compressor using two full adders.

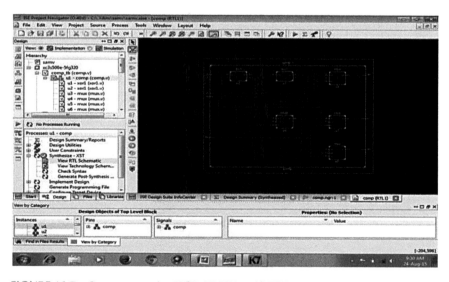

FIGURE 16.7   Compressor using XOR–XNOR and MUX.

Table 16.1 shows the comparison of 4-2 compressor in both using FA and using XNOR and MUX in aspect of power consumption, area, and delay.[6]

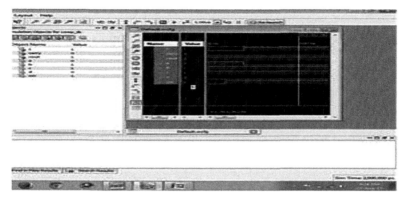

FIGURE 16.8   Schematic of 4-2 compressor using XOR–XNOR and MUX.

TABLE 16.1   Performance Analysis.

Power consumption	Using full adder	Using XNOR and MUX
	0.088 pW	**0.081 pW**
Area (used LUT)	2 μm	1 μm
Delay	6.837 ns	5.776 ns

16.5   CONCLUSION

A 4-2 compressor circuit based on a new XOR–XNOR designed provides better performance. The proposed XOR–XNOR design shows power consumption. The XOR provides maximum output delay and XNOR shows delay.[9] The performance of this circuit has been compared to earlier reported circuits in terms of power consumption[2] and maximum output delay. The proposed circuit result shows better performance than existing circuits in all aspects.[3]

KEYWORDS

- **full adder (FA)**
- **XOR–XNOR**
- **MUX**
- **Xilinx 10.1 ISE**
- **VERILOG**
- **MODEL SIM 6.3**

REFERENCES

1. Liang, J.; Han, J.; Lombardi, F. New Metrics for the Reliability of Approximate and Probabilistic Adders. *IEEE Trans. Comput.* **2013,** *63* (9), 1760–1771; Chakrapani, L. N.; A Probabilistic CMOS Switch and Its Realization by Exploiting Noise. In *Proc. IFIP-VLSI*, SoC, Perth, Western Australia, Oct. 2005.

2. Mahdiani, H. R.; Ahmadi, A.; Fakhraie, S. M.; Lucas, C. Bio-Inspired Imprecise Computational Blocks for Efficient VLSI Implementation of Soft-computing Applications. *IEEE Trans. Circuits Syst. I: Reg. Pap.* **2010,** *57* (4), 850–862.

3. Schulte, M. J.; Swartzlander, Jr., E. E. Truncated Multiplication with Correction Constant. In *VLSI Signal Process in VI*, 1993; pp. 388–396.

4. King, E. J.; Swartzlander, Jr., E. E. In *Data Dependent Truncated Scheme for Parallel Multiplication*, Proceeding of Thirty First Asilomar Conference on Signals, Circuits and Systems 1998; pp. 1178–1182.

5. Gupta, V.; Mohapatra, D.; Park, S. P.; Raghunathan, A.; Roy, K. IMPACT: M Precise Adders for Low-power Approximate Computing. In *Low Power Electronics and Design (ISLPED) 2011 International Symposium*, Aug 1–3, 2011.

6. Cheemalavagu, S.; Korkmaz, P.; Palem, V.; Akgul, B. E. S. Thirty First Asilomar Conference on Signals, Circuits and Systems, 1998; pp 1178–1182.

7. Kulkarni, P.; Gupta, P.; Ercegovac, M. D. Trading Accuracy for Power in a Multiplier Architecture. *J. Low Power Electr.* **2011,** *7* (4), 490–501.

8. Chang, C.; Gu, J.; Zhang, M. Ultra Low-Voltage Low-Power CMOS 4-2 and 5-2 Compressors for Fast Arithmetic Circuits. *IEEE Trans. Circuits Syst.* **2004,** *51* (10), 1985–1997.

9. Radhakrishnan, D.; Preethy, A. P. Low-Power CMOS Pass Logic 4-2 Compressor for High-speed Multiplication. *Proc. 43rd IEEE Midwest Symp. Circuits Syst.* **2000,** *3*, 1296–1298.

10. Wang, Z.; Jullien, G. A.; Miller, W. C. A New Design Technique for Column Compression Multipliers. *IEEE Trans. Comput.* **1995,** *44*, 962–970.

CHAPTER 17

IMPLEMENTATION OF A NEW VLSI ARCHITECTURE FOR ADD–MULTIPLY OPERATORS USING A MODIFIED BOOTH RECODING TECHNIQUE

K. NUNNY PRAISY* and K. S. N. RAJU

SVECW (Autonomous), Bhimavaram, Andhra Pradesh, India

Corresponding author. E-mail: knpraisy@gmail.com

ABSTRACT

The alternate operations are used to modify the applications in the digital signal processors. Mainly we will design the modified booth recoder with the sum of two numbers to increase their performance. The main operation that is performed in this technique is multiply and Accumulator (MAC) which is used to increase the performance in digital signal processor. This is mainly done with Fused Add Multiply (FAM) to reduce the critical path, power dissipation and area. This is introduced for the performance of the carry look-ahead adders (CLA) to reduce the Area-delay of the design with reversible logic gates.

17.1 INTRODUCTION

The main aspect of this design is to compress the complex arithmetic circuits. The base of the design consists of adders and multipliers. The digital signal processing (DSP) application can perform the large number of operations in this architecture.[1,2] The performance is mainly obtained in DSP applications and the allocation of the arithmetic operations with sign bit. This is to improve its performance and design of the arithmetic units in this architecture. The operation of adders and multipliers is taken with the help of multiply accumulator and multiply–add[3] to increase the area and critical path delay, by first taking the adder input to drive the output of the multiplier. This can be performed with the direct recoding of the sum of two numbers in its MB form to implement this technique.[4-6] The conventional circuits are mainly used for the combination of the circuits in gate level. The proposed technique is used to reduce its area, power dissipation, and critical path delay by using this S-MB structure with the alternate booth form with signed and unsigned digits with the conventional bit adders of full adder (FA) and half adder (HA).

17.2   MODIFIED BOOTH RECODING

Booth's algorithm gives a pair of bits of the N-bit multiplier in two's complement representation. This includes the least significant bits (LSB). For each bit, i running from 0 to $N - 1$, there exists the bits Y_i and Y_{i-1} when these two bits are equal and the product accumulator P is left unchanged. The final value of P is the signed product.[7,8] The multiplicand and product are not specified; typically, these are in two's compliment representation. It proceeds from LSB to most significant bit (MSB); starting at $i = 0$, the multiplication by 2^i is then replaced by incremental shifting of the P accumulator to the right.

The basic block diagram of the AM operator is to introduce architecture of the multiplier (Fig. 17.1).[1]

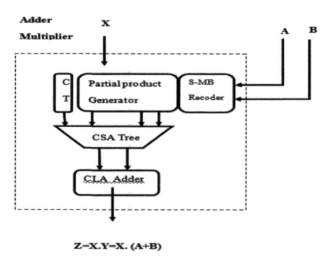

$$Z = X.Y = X. (A+B)$$

FIGURE 17.1   S-MB representation of sum of two numbers A and B.

This technique is mainly coded with the sum of two numbers in radix-4 representation.

$$Y_j^{MB} = -2_{y2j+1} + y_{2j} + y_{2j-1} \qquad (17.1)$$

In this, we can assume the multipliers of 2's compliments with this technique. In this, we are using three techniques S-MB-1, S-MB-2, and S-MB-3 with even and odd bits.

Gate level implementation is shown in Figure 17.2 and the partial products that are used to produce k's partial products are shown in Figure 17.3.

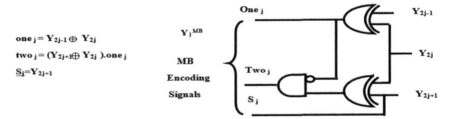

one$_j$ = Y$_{2j-1}$ ⊕ Y$_{2j}$

two$_j$ = (Y$_{2j+1}$⊕ Y$_{2j}$).one$_j$

S$_j$=Y$_{2j+1}$

FIGURE 17.2 Gate level implementation.

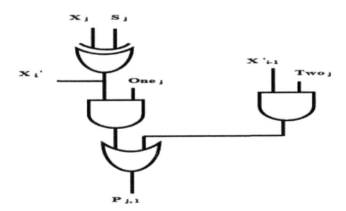

FIGURE 17.3 Generation of bits.

To generate the partial products, we use the carry save adder Wallace tree with the correction term as

$$Z = X \cdot Y = CT + \sum_{j=0}^{k-1} pp_j 2^{2j} \qquad (17.2)$$

Table 17.1 represents the encoding table with three bits showing sign, one$_j$, and two$_j$.

TABLE 17.1 MB Encoding Table.

Binary			Y_j^{MB}	MB encoding			Input carry
Y_{2j+1}	Y_{2j}	Y_{2j-1}		Sign = s_j	×1 = 1$_j$	×2 = 2$_j$	$C_{in\,j}$
0	0	0	0+1	0	0	0	0
0	0	1	+1	0	1	0	0
0	1	0	+2	0	1	0	0
0	1	1	−2	0	0	1	0
1	0	0	−1	1	0	1	1
1	0	1	−1	1	1	0	1
1	1	0	0	1	1	0	1
1	1	1		1	0	0	0

In this technique, we use the sign bits to perform with HAs and FAs (Fig. 17.4).[3] Here, we will use two sign HAs as HA* and HA** and the Boolean equations and Tables 17.2–17.5[3] are shown below.

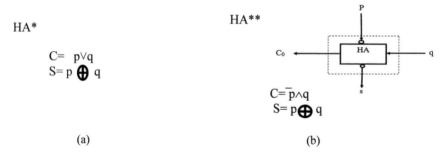

HA*

$$C = p \vee q$$
$$S = p \oplus q$$

HA**

$$C = \bar{p} \wedge q$$
$$S = p \oplus q$$

(a)   (b)

FIGURE 17.4 Boolean equations for (a) HA* and (b) HA**.

TABLE 17.2 HA* Basic Operation.

Inputs		Output value[1]	Outputs	
p(+)	q(+)		c(+)	s(−)
0	0	0	0	0
0	1	+1	1	1
1	0	+1	1	1
1	1	+2	1	0

Output value = −2c + s = −p − q

TABLE 17.3   HA* Dual Operation.

Inputs		Output value[2]	Outputs	
p(−)	q(−)		c(−)	s(+)
0	0	0	0	0
0	1	−1	1	1
1	0	−1	1	1
1	1	−2	1	0

Output value = $2c - s = p + q$

TABLE 17.4   HA** Operation.

Inputs		Output value[3]	Outputs	
p(+)	q(+)		c(+)	s(−)
0	0	0	0	0
0	1	+1	1	1
1	0	−1	0	1
1	1	0	0	0

Output value = $2c - s = -p + q$

TABLE 17.5   FA* Operation.

Inputs			Output value[2]	Outputs	
p(−)	q(−)	$C_i(+)$		$C_0(+)$	s(+)
0	0	0	0	0	0
0	0	1	+1	0	1
0	1	0	−1	1	1
0	1	1	0	0	0
1	0	0	−1	1	1
1	0	1	0	0	0
1	1	0	−2	1	0
1	1	1	−1	1	1

Output value = $-2c_0 + s = -p - q + c_i$

17.3   BOOTH RECODING TECHNIQUES

17.3.1   S-MB1 TECHNIQUE

In this technique, we use two consecutive bits A and B with A (a_{2j}, a_{2j+1}) and B (b_{2j}, b_{2j+1}) (Fig. 17.5). We use FAs and HAs in this technique with conventional output as FA*. In this, we are using odd bit width output value as FA** (Fig. 17.6).[4]

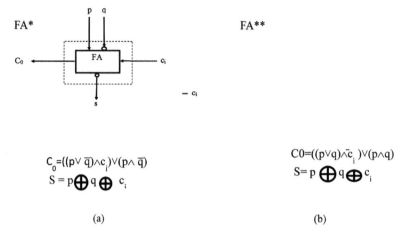

$$C_0 = ((p \vee \bar{q}) \wedge c_i) \vee (p \wedge \bar{q})$$
$$S = p \oplus q \oplus c_i$$

$$CO = ((p \vee q) \wedge \bar{c}_i) \vee (p \wedge q)$$
$$S = p \oplus q \oplus c_i$$

(a)   (b)

FIGURE 17.5   Boolean equations for (a) FA* and (b) FA**.

$$FA^* = 2c_0 - S = p - q + c_i \qquad\qquad FA^{**} = -2c_0 + S = -p - q + c_i$$

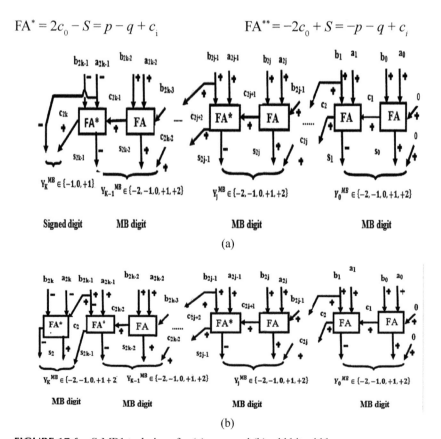

FIGURE 17.6   S-MB1 technique for (a) even and (b) odd bit width.

17.3.2   S-MB2 TECHNIQUE

The second technique is used to describe the even and odd bit width with consecutive bits as FA to produce the sum and carry as s_{2j}, c_{2j+1} and the inputs of FA are a_{2j}, b_{2j}, $c_{2j,1}$. In this, the output sum s_{2j} is driven by the carry c_{2j+1} and the inputs of HA with a_{2j}, b_{2j}. The output value of this technique is HA* (Fig. 17.7).[5]

$$\text{HA}^* = -2C + S = -p - q$$

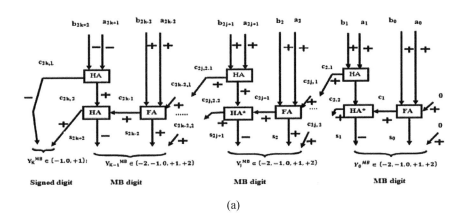

(a)

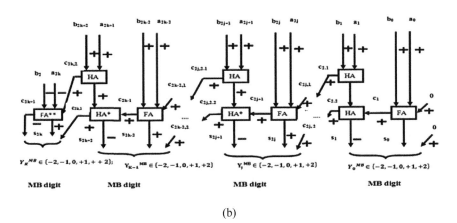

(b)

FIGURE 17.7   S-MB2 technique for (a) even and (b) odd bit width.

17.3.3 S-MB3 TECHNIQUE

In this technique, we introduce the S-MB3 architecture, where the sum s_{2j} and carry c_{2j+1} are to produce FA, so the $c_{2j,1}$ is used as output carry of HA* and inputs as a_{2j}, b_{2j}, the output value as HA** (Fig. 17.8).[5]

$$HA^{**} = 2c - s = -p + q$$

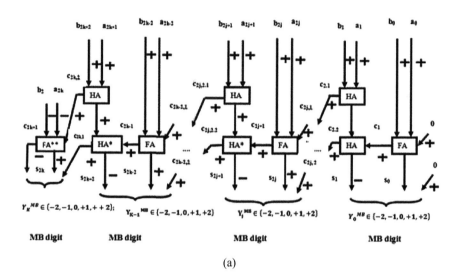

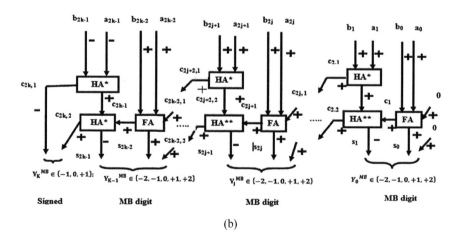

FIGURE 17.8 S-MB3 technique for (a) even and (b) odd bit width.

17.3.4 UNSIGNED INPUT NUMBERS

In this technique, we are using input bits both A and B as unsigned numbers and also using the S-MB schemes as even and odd, regarding whether the input as A and B in the sign MSB (Figs. 17.9–17.13).[6]

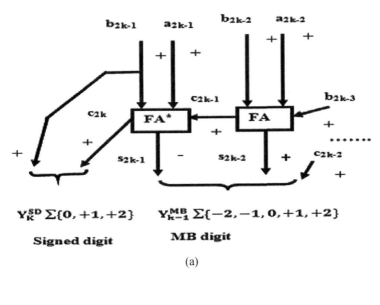

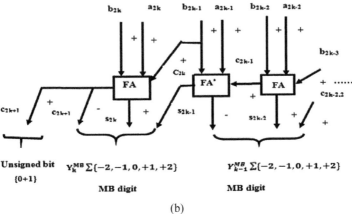

FIGURE 17.9 Unsigned input technique S-MB1 for (a) even and (b) odd.

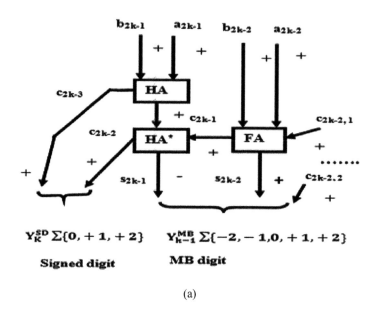

(a)

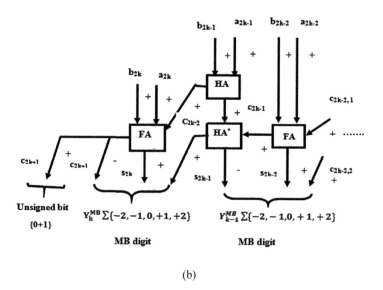

(b)

FIGURE 17.10   Unsigned input numbers S-MB2 for (a) even and (b) odd.

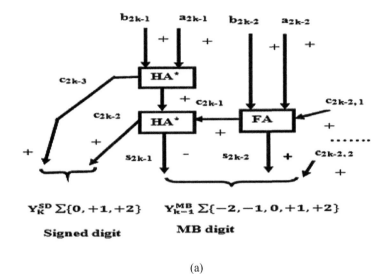

(a)

(b)

FIGURE 17.11   Unsigned input numbers S-MB3 for (a) even and (b) odd.

17.4   RESULTS AND COMPARISONS

17.4.1   RTL SCHEMATIC

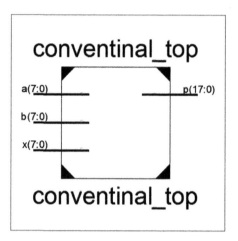

FIGURE 17.12   Conventional register-transfer level (RTL) schematic.

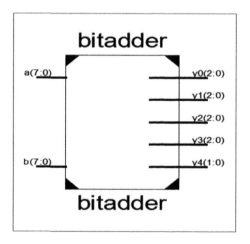

FIGURE 17.13   Bit adder register-transfer level (RTL) schematic.

17.4.2   S-MB1 RECODING TECHNIQUE

This is the output waveform of signed S-MB1 recoding technique for both even and odd bit width (Fig. 17.14).[5]

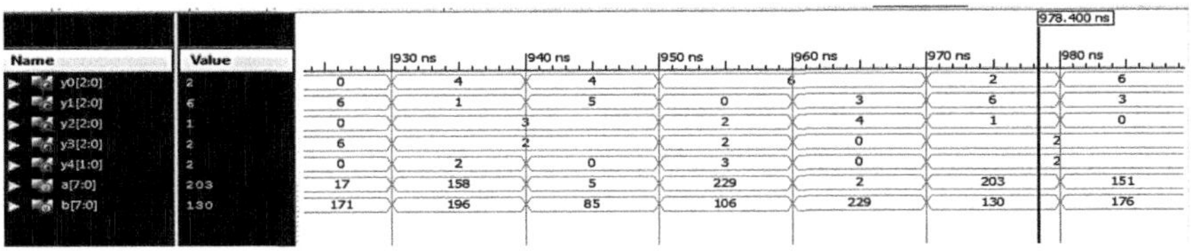

(a)

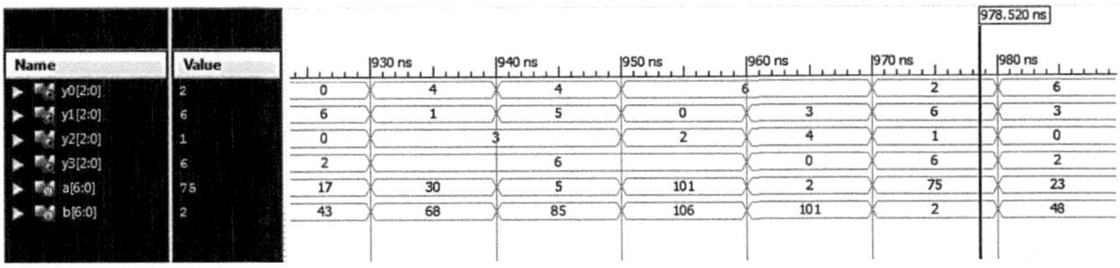

(b)

FIGURE 17.14 S-MB1 recoding technique for (a) even and (b) odd.

17.4.3 S-MB2 RECODING TECHNIQUE

This is the output waveform of signed S-MB2 recoding technique for both even and odd bit width (Fig. 17.15).[5]

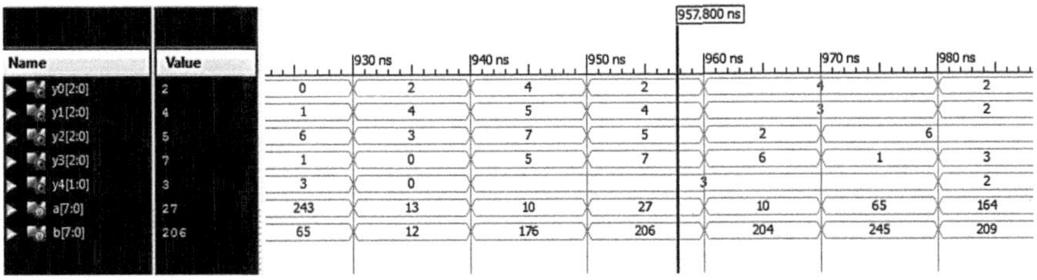

(a)

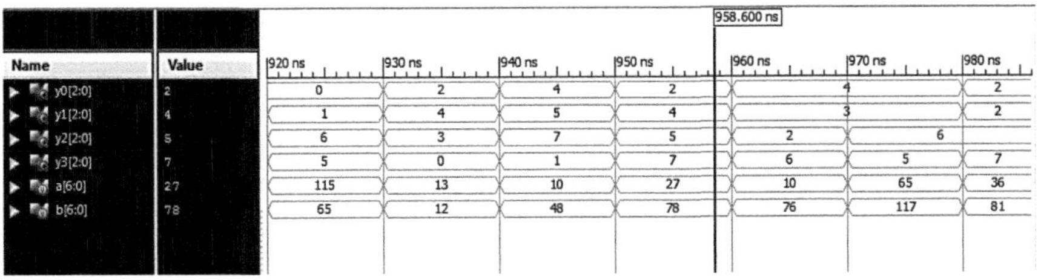

(b)

FIGURE 17.15 S-MB2 recoding technique for (a) even and (b) odd.

17.4.4   S-MB3 RECODING TECHNIQUE

This is the output waveform of signed S-MB3 recoding technique for both even and odd bit width (Fig. 17.16).[6]

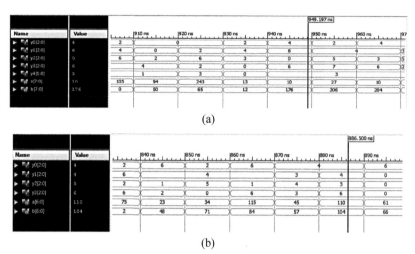

(a)

(b)

FIGURE 17.16   S-MB3 recoding technique for (a) even and (b) odd.

17.4.5   UNSIGNED INPUT NUMBERS OF S-MB1 RECODING TECHNIQUE

This is the output waveform for unsigned input numbers of S-MB1 recoding technique for both even and odd bit width (Fig. 17.17).[6]

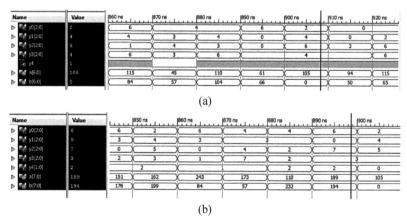

(a)

(b)

FIGURE 17.17   Unsigned input numbers of S-MB3 recoding technique for (a) even and (b) odd.

17.4.6 UNSIGNED INPUT NUMBERS OF S-MB2 RECODING TECHNIQUE

This is the output waveform for unsigned input numbers of S-MB2 recoding technique for both even and odd bit width (Fig. 17.18).[7]

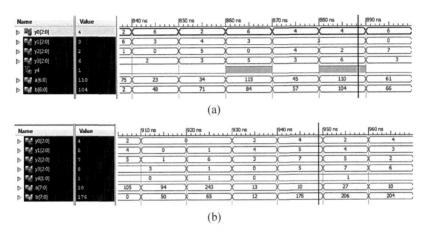

(a)

(b)

FIGURE 17.18 Unsigned input numbers of S-MB3 recoding technique for (a) even and (b) odd.

17.4.7 UNSIGNED INPUT NUMBERS OF S-MB3 RECODING TECHNIQUE

This is the output waveform for unsigned input numbers of S-MB3 recoding technique for both even and odd bit width (Fig. 17.19).[7]

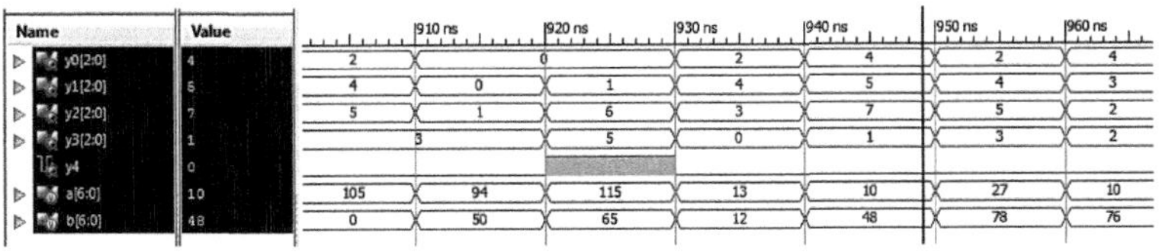

(a)

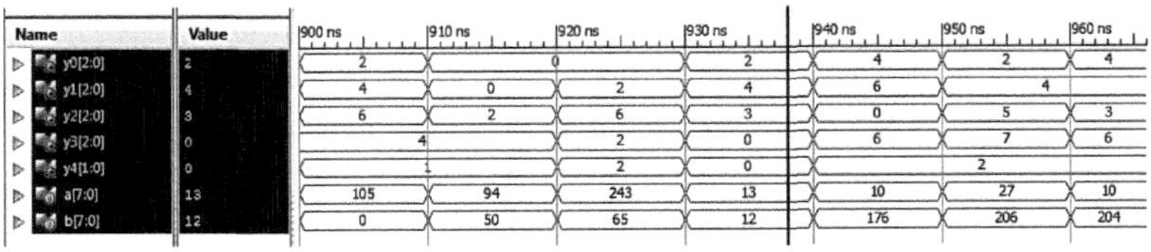

(b)

FIGURE 17.19 Unsigned input numbers of S-MB3 recoding technique for (a) even and (b) odd.

17.5   COMPARISONS

In this, we compare the values of area and delay with three different techniques (Tables 17.6 and 17.7).[8]

TABLE 17.6   Comparison of Area for the Three S-MB Techniques.

Parameters	S-MB1 technique		S-MB2 technique		S-MB3 technique	
	Even	Odd	Even	Odd	Even	Odd
Number of slice LUTs	97	77	100	81	100	81
Number of occupied slices	54	49	58	44	58	44
Number of bonded IOBs	42	38	42	38	42	38

TABLE 17.7   Comparison of Delay for the Three S-MB Techniques.

Parameters	S-MB1 technique		S-MB2 technique		S-MB3 technique	
	Even	Odd	Even	Odd	Even	Odd
Delay	8.690 ns	7.414 ns	9.640 ns	7.382 ns	9.537 ns	7.280 ns

17.6   CONCLUSION

In this chapter, we simulate the result of booth recoder using S-MB recoding technique. This design consists of three recoding techniques, S-MB1, S-MB2, and S-MB3. The S-MB techniques are used for both signed and unsigned bits with two different cases. It is also used for even and odd bit widths. The main purpose of the proposed technique with the existing technique is to reduce the area consumption, critical path delay, and power consumption with more improvements in its performance.

KEYWORDS

- sum to modified booth
- CLA adder
- alternate booth recoder
- VLSI architecture
- full adder
- half adder

REFERENCES

1. Amaricai, A.; Vladutiu, M.; Boncalo, O. Designissues and Implementations for Floating-point Divide-add Fused. *IEEE Trans. Circuits Syst. II–Exp. Briefs* **2010,** *57* (4), 295–299.
2. Swartzlander, E. E.; Saleh, H. H. M. FFT Implementation with Fused Floating-point Operations. *IEEE Trans. Comput.* **2012,** *61* (2), 284–288.
3. Kwon, O.; Nowka, K.; Swartzlander, E. E. A 16-Bit by 16-Bit MAC Design Using Fast 5:3 Compressor Cells. *J. VLSI Signal Process. Syst.* **2002,** *31* (2), 77–89.
4. Chen, L.-H.; Chen, O. T.-C.; Wang, T.-Y.; Ma, Y.-C. A Multiplication–Accumulation Computation Unit with Optimized Compressors and Minimized Switching Activities. In *Proc. IEEE Int., Symp. Circuits Syst.* **2005,** *6,* 6118–6121.
5. Seo, Y.-H.; Kim, D.-W. A New VLSI Architecture of Parallel Multiplier–Accumulator Based on Radix-2 Modified Booth Algorithm. *IEEE Trans. Very Large Scale Integr. (VLSI) Syst.* **2010,** *18* (2), 201–208.
6. Zimmermann, R.; Tran, D. Q. Optimized Synthesis of Sum-of-Products. In *Proc. Asilomar Conf. Signals, Syst. Comput.*, Pacific Grove, Washington, DC, 2003; pp 867–872.
7. Daumas, M.; Matula, D. W. A Booth Multiplier Accepting Both a Redundant and a Nonredundant Input with No Additional Delay. In *Proc. IEEE Int. Conf. Appl.-Specif. Syst., Architect., Processors*, 2000; pp 205–214.
8. Huang, Z.; Ercegovac, M. D. High-performance Low-power Left to Right Array Multiplier Design. *IEEE Trans. Comput.* **2005,** *54* (3), 272–283.

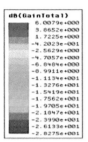

FIGURE 1.12   Gain plot of the two-element antenna array of mitered bend feed network.

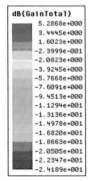

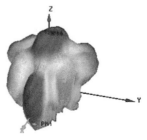

FIGURE 2.11   Gain plot of the two-element antenna array of quarter-wave feed network.

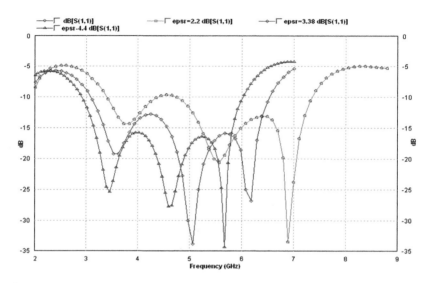

FIGURE 3.10   Measured return loss for the proposed antenna with various dielectric constants. **Note:** Other parameters are the same as in Figure 3.2.

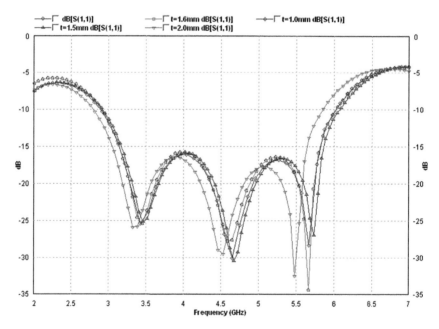

FIGURE 3.11   Measured return loss for the proposed antenna with various substrate thicknesses.

Note: Other parameters are the same as in Figure 3.2.

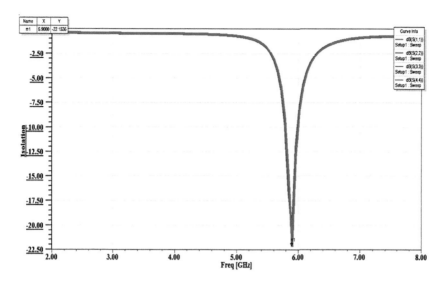

FIGURE 4.4   Reflection coefficient of the conventional MIMO antenna.

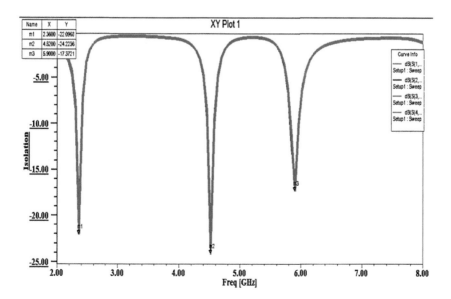

FIGURE 4.5   Reflection coefficient of the proposed octagon split-ring MIMO antenna.

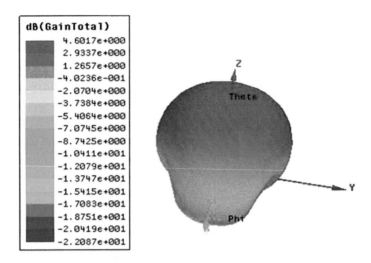

FIGURE 4.6   3D plot of conventional MIMO antenna gain.

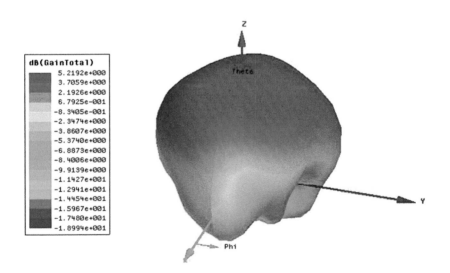

FIGURE 4.7   3D plot of octagon split-ring MIMO antenna gain.

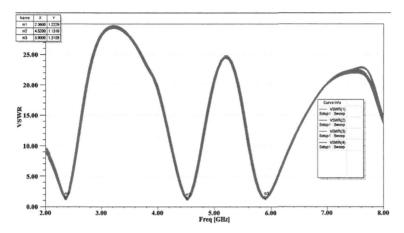

FIGURE 4.8   VSWR of octagon split-ring MIMO antenna.

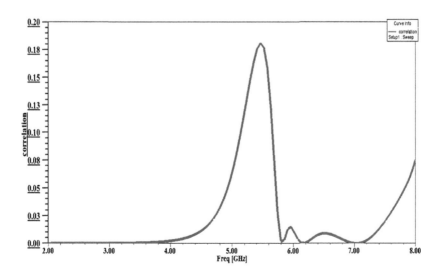

FIGURE 4.9   Correlation coefficient of octagon split-ring MIMO antenna.

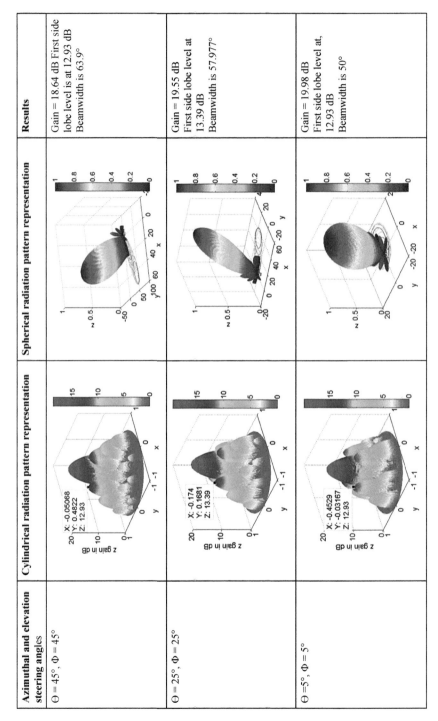

FIGURE 7.3 Phased array antenna with different steering angles. Adapted from Ref. [8].

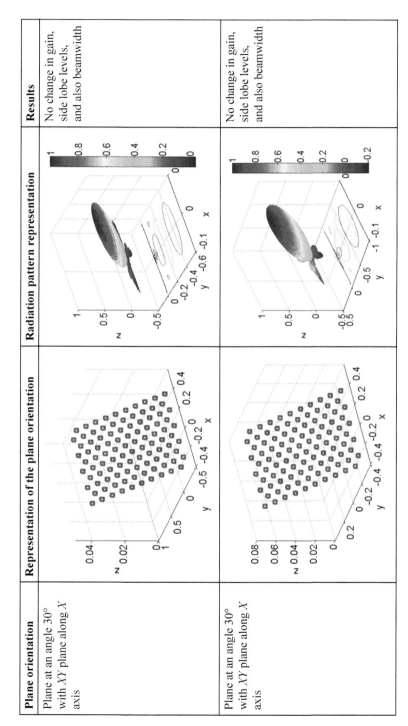

FIGURE 7.4 The radiation pattern of phased array antenna for different steering angles. Adapted from Ref. [8].

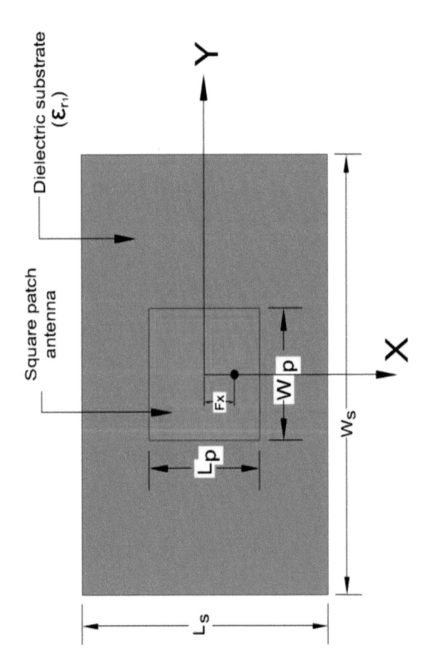

FIGURE 8.2   Geometry of square-patch antenna (top view).

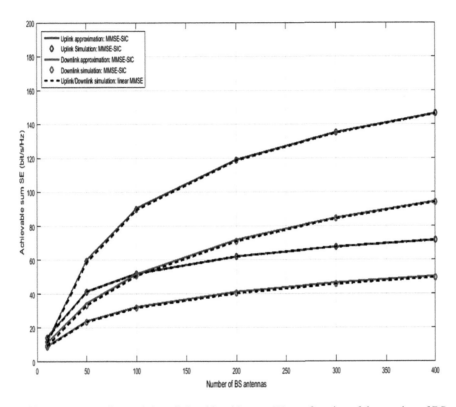

FIGURE 10.1 Uplink and downlink achievable sum SE as a function of the number of BS antennas for $K = 10$.

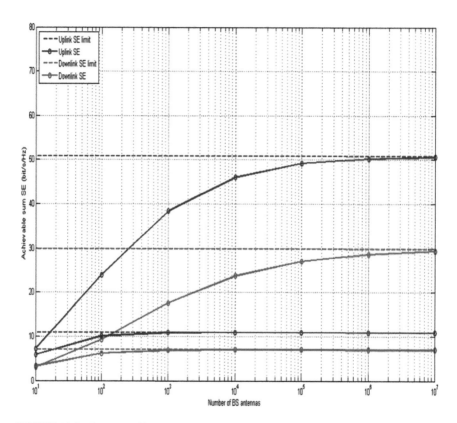

FIGURE 10.2 Power scaling law for $K = 10$, $N = 3$, $a_r = 0$, and $a_t = 0.4$.

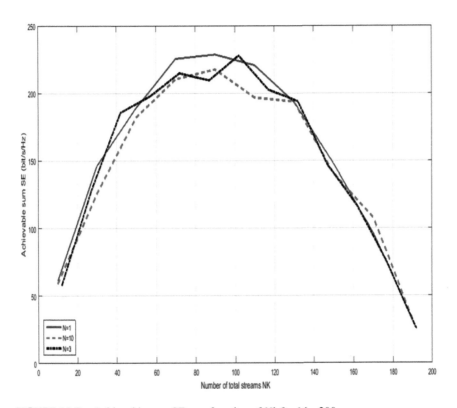

FIGURE 10.3   Achievable sum SE as a function of Nk for $M = 200$.

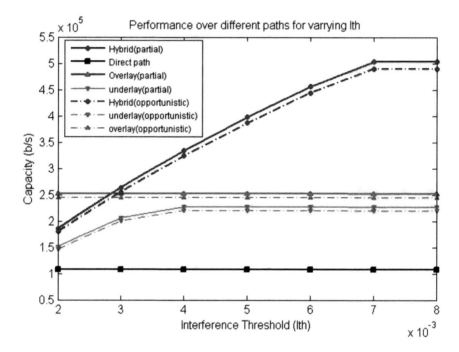

FIGURE 12.3   Capacity over different paths with partial and opportunistic relay selection and fixed PU transmitting power.

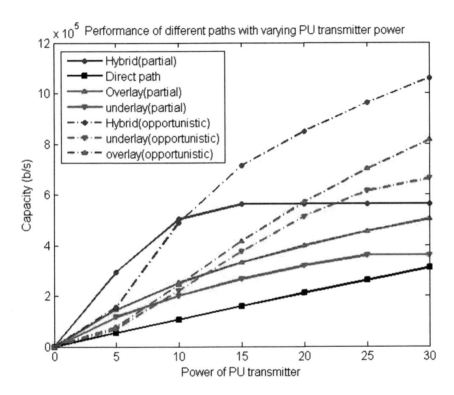

FIGURE 12.4   Capacity over different paths with partial and opportunistic relay selection and fixed interference threshold.

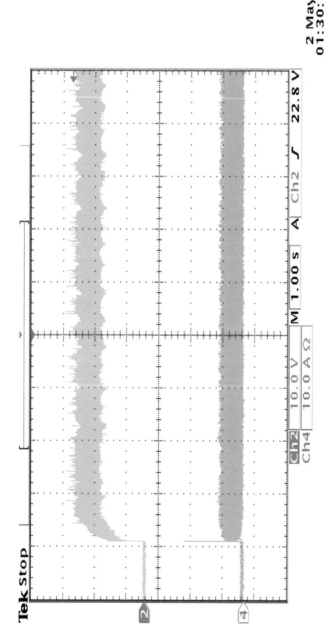

FIGURE 22.10 The oscillation of the voltage and current waveform.

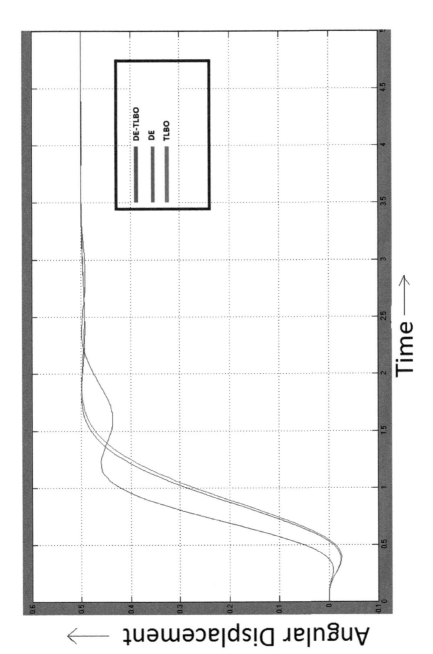

FIGURE 28.7 Comparison of different algorithms.

CHAPTER 18

DESIGN OF A BAUGH–WOOLEY MULTIPLIER IN QUANTUM DOT CELLULAR AUTOMATA USING AN AREA OPTIMIZED FULL ADDER

B. RAMESH[1*] and M. ASHA RANI[2]

[1]Department of Electronics and Communication Engineering, Kamala Institute of Technology & Science, Huzurabad 505468, Telangana, India

[2]Department of Electronics and Communication Engineering, JNTUH College of Engineering, Hyderabad 500085, Telangana, India

*Corresponding author. E-mail: Brameshb2@rediffmail.com

ABSTRACT

The Quantum dot Cellular Automata (QCA) is an emerging nanotechnology that has significant attractive characteristics such as smaller size, lower power consumption, and fast speed than conventional transistor-based complementary metal-oxide-semiconductor technology. To explore the features of QCA technology, the digital circuit designs have been investigated. The arithmetic, logic, and memory circuit designs have been most interesting area of research from the last decade. This paper presents an area-efficient QCA Baugh-Wooly multiplier design using a novel area optimized QCA full adder. The design proposed is simulated in QCA designer; the results obtained confirm that the QCA layout area is reduced by 11–78%. Further, the cell count is reduced by 13–49% with optimum latency in comparison with existing multiplier designs.

18.1   INTRODUCTION

The nanoelectronic technology has been the emerging area to the research community since the last decade because the nanoelectronic devices have an upcoming alternative to the conventional complementary metal-oxide-semiconductor (CMOS) based transistors in very-large-scale integrationtechnology, to overcome the challenges in CMOS technology at nanometer scales. To increase the density of digital systems for portable, low-power designs, the traditional CMOS technology is playing vital role for the past few decades. But for the past few years at nanometer scale, CMOS technology is facing new challenges, which are short channel effects like subthreshold leakage currents, drain-induced barrier lowering, hot carrier effects, velocity saturation, punch-through, etc.[1] To overcome CMOS design challenges, in the year 2007, ITRS has identified few nanoelectronic technologies, like resonant tunneling diode, quantum dot cellular automata (QCA), carbon nanotubes, single electron transistor, etc., to replace CMOS transistor-based technology. In all these technologies, QCA seems to be an attractive alternative technology for the CMOS technology with its similarity in top-down approach.[2]

Lent[3] in 1993 introduced QCA technology as an attractive alternative to replace the conventional CMOS technology. In recent years, QCA has gained a lot of popularity due to its computing features for logic functions at nanometer scale. QCA technology provides very high density, high operational frequency (THz range)[4,5] and very low power consumption.[6] In this chapter, a novel area-optimized, high-performance QCA Baugh–Wooley multiplier is presented. It takes the advantage of area-efficient QCA full adder (FA)[14] in the implementation of addition operation in multiplication. QCA layout of a 4-bit multiplier is designed and simulated in QCADesigner.[7] The results of proposed design are compared with its existing counterparts[8–11] for the QCA design metrics cell count, area, and latency (number of clock cycles).[12]

The rest this chapter is organized as follows: brief introduction of QCA, clocking schemes, and QCA wire crossovers are covered in Section 18.2. A brief report on related previous work is presented in Section 18.3. The design of proposed efficient Baugh–Wooley multiplier is presented in Section 18.4. The results and comparison of design metrics of proposed with existing designs are presented in Section 18.5. Finally, conclusions are presented in Section 18.6.

18.2 QCA BASICS

18.2.1 CELL, WIRE, AND GATES

The logic states in QCA technology are based on the individual electron location, instead of voltage levels in conventional CMOS technology. The basic component in QCA technology is a quantum dot of a 5-nm diameter single electron container, and primitive logic element in QCA is a 18-nm × 18-nm square nanostructure called QCA cell, which consists of four quantum dots at the corners of cell; there is 5 nm spacing between the dots and 20 nm spacing between the centers of two adjacent cells. A QCA cell contains two electrons and two empty dots and always electrons reside in diagonally opposite dots due to columbic repulsion. The polarization of a QCA cell defined based on the position of electrons in the cell (Fig. 18.1); the polarization $P = -1$ represents binary 0 and the $P = +1$ represents binary 1.[13] The electrons in a QCA cell can only tunnel between dots inside a cell but not between cells due to high barriers between cells.

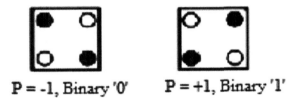

P = -1, Binary '0'   **P = +1, Binary '1'**

FIGURE 18.1   QCA cells, polarization $P = -1$ represent binary 0 and $P = +1$, binary 1.

A wire in QCA is an array of cells (Fig. 18.2) to transmit signal from one cell to the next cell. An inverter and majority gate are basic logic computing elements in QCA. There are two types of inverters[14] in QCA designs; one using four cells and second one using seven cells (Fig. 18.3) complement the given input. A three input majority gate consists of five cells and produces majority logic output based on the three binary inputs (Fig. 18.4). The logic function of majority gate is $Y = M(A, B, C) = AB + BC + CA$ for the binary inputs A, B, C. For $C = 0$, it becomes AND gate, and for $C = 1$, it becomes OR gate.

input=1 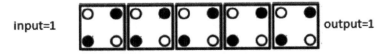 output=1

FIGURE 18.2   QCA wire.

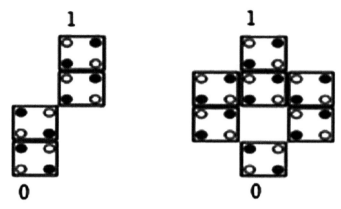

FIGURE 18.3   QCA inverters.

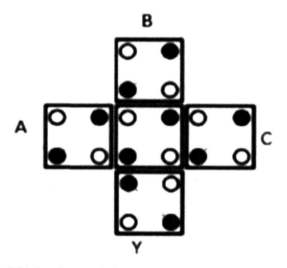

FIGURE 18.4   QCA three input majority gate.

18.2.2 CLOCKING SCHEMES IN QCA

Clock signals in QCA design are used to excite the QCA cells; there are four clocking phases—switch, hold, release, and relax[15] (Fig. 18.5). Each phase has a shift of 90°. In clock switch phase, the cell begins with unpolarized low potential barriers, and the barriers rise to high during this phase. During the clock hold phase, the potential barriers held at high, during the release phase, the potential barriers are lowered, and finally during the relax phase, the barriers remain at low and keep the cell in an unpolarized state.

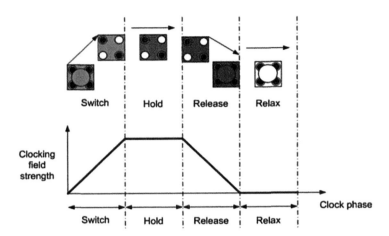

FIGURE 18.5   Clocking phases in QCA.

18.2.3   WIRE CROSSOVERS IN QCA

Two wire crossovers are used in QCA designs,[12] one is coplanar wire crossover and second one is multilayer wire crossover. The coplanar wire crossovers are designed using four clock zones for QCA cells, and they are clock 0, clock 1, clock 2, and clock 3, each clock zone has a phase difference of 90°. The QCA cells with clock 0 and clock 2 are phase shifted by 180° and the intersection of these two types of cells can form a coplanar crossover. Similarly, a QCA cell with clock 1 and clock 3 are shifted by 180° and these two types can form a coplanar wire crossover (Fig. 18.6). A multilayer wire crossover is designed using four QCA layers (Fig. 18.7). The type of wire crossovers used in circuits is one of the major QCA design metrics, because the fabrication cost of multilayer wire crossover is three times of the cost of coplanar wire crossover.[12]

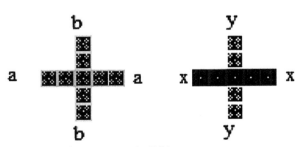

FIGURE 18.6   Coplanar wire crossovers in QCA.

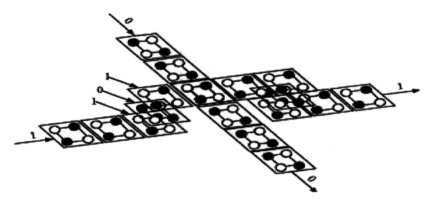

FIGURE 18.7   Multilayer wire crossover in QCA.

18.3   RELATED WORK

The QCA multipliers designed earlier were presented in Refs. [8–11]. A bit-serial multiplier design in Ref. [8] used carry flow adder and carry look-ahead adder for addition operation in the multiplication; its delay is high because it is serial adder. A serial-parallel multiplier presented in Ref. [9] used bit-serial adder for the implementation, so the delay of design is very high. The Wallace and Dadda multiplier presented in Ref. [10] is more complex and high delayed, and a Baugh–Wooley multiplier presented in Ref. [11] used multilayer wire crossover-based adders for the implementation of addition operation for multiplication; its complexity is very high in terms of QCA design metrics. In this chapter, an area-efficient multiplier is presented, which uses an area-optimized QCA FA[14] for the addition operation in multiplication. The coplanar wire crossover-based FA in this multiplier reduces the circuit complexity. The designed QCA layout is simulated in QCADesigner;[7] results obtained from the QCA implementation are compared with its existing counterparts.

18.3.1   PROPOSED BAUGH–WOOLEY MULTIPLIER IN QCA

The multiplication in Baugh–Wooley approach is performed for two's complement numbers A and B ($A_3A_2A_1A_0$ and $B_3B_2B_1B_0$, respectively), and the product is given by $P_7P_6P_5P_4P_3P_2P_1P_0$ (Fig. 18.8).

The partial products of the multiplication (Fig. 18.8) are implemented using an AND gate form of the three input majority gate, and an area-optimized

QCA FA[14] is used to implement the FA blocks in a Baugh–Wooley multiplier (Fig. 18.9). The area and QCA cell count of the proposed multiplier are reduced with optimum circuit delay.

					A_3	A_2	A_1	A_0
			X		B_3	B_2	B_1	B_0
				1	$\overline{A_3B_0}$	A_2B_0	A_1B_0	A_0B_0
					$\overline{A_3B_1}$	A_2B_1	A_1B_1	A_0B_1
				$\overline{A_3B_2}$	A_2B_2	A_1B_3	A_0B_2	
1		A_3B_3	$\overline{A_2B_3}$	$\overline{A_1B_3}$	$\overline{A_0B_3}$			
P_7	P_6	P_5	P_4	P_3	P_2	P_1	P_0	

FIGURE 18.8   4-bit Baugh–Wooley multiplication.

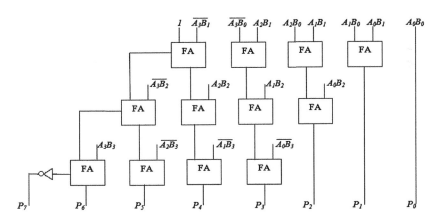

FIGURE 18.9   Circuit diagram of 4-bit Baugh–Wooley multiplier.

The block diagram of a novel area-optimized QCA FA and its QCA layout is designed in QCADesigner[7] using three majority gates, two inverters, and two wire crossovers (Fig. 18.10). Logic functions expression in majority gate functions of a FA carry and sum are given in eqs 18.1 and 18.2, respectively. To reduce the QCA cost function, coplanar crossovers are used instead of multilayer crossovers.[12] A FA for Baugh–Wooley multiplier presented in Ref. [11] was implemented with 78 QCA cells, 0.06 μm² area, a novel FA for multiplier designed only with 52 QCA cells, and 0.38 μm² area to optimize the multiplier.

$$C_{out} = M(a,b,c) \tag{18.1}$$

$$S = M\left(C_{out}^{1}, M(a,b,c^{1}), c\right) \tag{18.2}$$

where a, b, and c are inputs and output carry is C_{out}, and output sum is S. Equations (18.1) and (18.2) can be expanded to find the logic functions of carry and sum of a FA as given below.

$$C_{out} = ab + bc + ca$$
$$S = \left(ab + bc + ca\right)^{1}\left\{\left(ab + bc^{1} + ac^{1}\right) + c\right\} + \left(ab + bc^{1} + ac^{1}\right)c$$
$$= \left(a^{1}b^{1} + b^{1}c^{1} + c^{1}a^{1}\right)\left(a + b + c\right) + abc$$
$$= a^{1}b^{1}c + ab^{1}c^{1} + a^{1}bc^{1} + abc = a \oplus b \oplus c$$

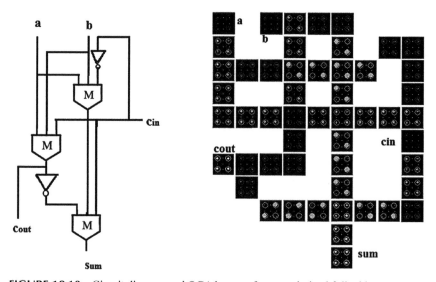

FIGURE 18.10   Circuit diagram and QCA layout of area optimized full adder.

The circuit diagram of proposed area-optimized QCA Baugh–Wooley multiplier with 12 FAs, and QCA layout design using an area-optimized QCA FA are shown in Figures 18.9 and 18.11, respectively. Results obtained from proposed design compared with previous designs are presented in the next section.

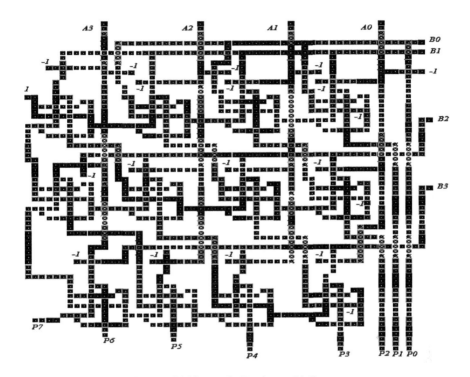

FIGURE 18.11   QCA layout of 4-bit Baugh–Wooley multiplier.

18.4   RESULTS AND COMPARISON

QCA layout of proposed Baugh–Wooley multiplier is designed using QCADesigner.[7] For the simulation here selected, QCA cell height and width are 18 nm, quantum dot diameter is 5 nm, and distance between centers of two adjacent cells is 20 nm. The simulation-type setup is chosen as vector table and *coherence vector engine* setup has been used for simulation.

The results of proposed design are compared with its existing counterparts for QCA design metrics as illustrated in Table 18.1. The obtained results confirm that the area of proposed design is reduced by 11–78%. Further, the QCA cell count is reduced by 13–49% with optimum number of clock cycles.

TABLE 18.1   Comparison of Results.

Adder	Cell count	Area (μm^2)	No. of clock cycles
4 × 4 Wallace[10]	3295	7.39	10
4 × 4 Dadda[10]	3384	7.51	12
4-bit Serial–parallel[8]	406	0.4935	8
4 × 4 Baugh–Wooley[11]	1982	1.8	4.75
Proposed	1726	1.6	4.75

18.5   CONCLUSION

In this chapter, design of a new area-efficient QCA-based Baugh–Wooley multiplier is presented. A 4 × 4 multiplier QCA layout is designed using a novel area-efficient QCA FA. The proposed multiplier layout simulated in QCADesigner, the area occupancy is up to 11% less than the best existing counterparts, and cell count is saved up to 13% in comparison to the best previous designs with an optimum clock delay. Further, the circuit complexity is reduced significantly by using coplanar wire crossovers in place multilayer wire crossovers.

KEYWORDS

- **nanoelectronic technology**
- **nanoelectronic devices**
- **CMOS technology**
- **quantum dot cellular automata**
- **carbon nanotubes**

REFERENCES

1. Kim, Y. B. Challenges for Nanoscale MOSFETs and Emerging Nanoelectronics. *IEEE Trans. Electr. Electron. Mater.* **2010,** *11* (3), 93–105.
2. International Technology Roadmap for Semiconductors (ITRS). http://www.itrs2.net, 2007.
3. Lent, C. S.; Tougaw, P. D.; Porod, W.; Bernstein, G. H. Quantum Cellular Automata. *Nanotechnology* **1993,** *4* (1), 49–57.

4. Lu, Y.; Liu, M.; Lent, C. S. Molecular Quantum-dot Cellular Automata: From Molecular Structure to Circuit Dynamics. *J. Appl. Phys.* **2007,** *102* (3), 034311-1–034311-7.

5. Lu, Y.; Liu, M.; Lent, C. S. Molecular Electronics—From Structures to Circuit Dynamics. In *Proc. 6th IEEE Conf. Nanotechnology,* July 2006; pp 62–65.

6. Blair, E. P.; Yost, E.; Lent, C. S. Power Dissipation in Clocking Wires for Clocked Molecular Quantum-dot Cellular Automata. *J. Comput. Electron.* **2010,** *9* (1), 49–55.

7. Walus, K.; Dysart, T.; Jullien, G. QCADesigner: A Rapid and Simulation Tool for Quantum-Dot Cellular Automata. *IEEE Trans. Nanotechnol.* **2004,** *3* (1), 26–29.

8. Cho, H.; Swartzlander, E. Adder and Multiplier Design in Quantum-dot Cellular Automata. *IEEE Trans. Comput.* **2009,** *58* (6), 721–727.

9. Walus, K.; Jullien, G.; Dimitrov, V. Computer Architecture Structures for Quantum Dot Cellular Automata. In *Record of Thirty-seventh Asilomar Conference on Signals, Systems and Computers*, 2003; pp 1435–1439.

10. Kim, S.; Swartzlander, E. Parallel Multiplier for Quantum Dot Cellular Automata. In *Proc. of IEEE Nanotechnology Materials and Devices Conference*, 2009; pp 68–72.

11. Pudi, V.; Sridharan, K. Efficient Design of Baugh–Wooly Multiplier in Quantum Dot Cellular Automata. In *Proc. of 13th IEEE Int. Conf. on Nanotechnology*, Beijing, China, August 5–6, 2013.

12. Liu, W.; Liang, L.; O'Neill, M.; Swartzlander, E. E. A First Step Towards Cost Functions for Quantum Dot Cellular Automata Designs. *IEEE Trans. Nanotechnol.* **2014,** *13* (3), 476–487.

13. Lent, C. S.; Tougaw, P. D. A Device Architecture for Computing with Quantum Dots. *Proc. IEEE* **1997,** *85* (4), 541–557.

14. Ramesh, B.; Asha Rani, M. Design of Binary to BCD Code Converter Using Area Optimized Quantum Dot Cellular Automata Full Adder. *Int. J. Eng.* **2015,** *9* (4), 49–64.

15. Walus, K.; Jullien, G. A. Design Tools for an Emerging SOC Technology: Quantum-dot Cellular Automata. *Proc. IEEE* **2006,** *94* (6), 1225–1244.

PART IV
Embedded Systems

CHAPTER 19

DESIGN AND ANALYSIS OF A HYBRID 4-2 APPROXIMATE COMPRESSOR FOR MULTIPLICATION

K. SATISHA* and M. V. GANESWARA RAO

Department of ECE, SVECW, Bhimavaram, Andhra Pradesh, India

Corresponding author. E-mail: kodi.satisha@gmail.com

ABSTRACT

In most digital signal processing (DSP) systems, one of the key hardware block is a multiplier. In a typical DSP applications, a multiplier plays an important role that includes digital communications, digital filtering, and spectral analysis. Many present that DSP applications are targeted at portable, battery-operated systems so that one of the primary design constraint is a power dissipation. In the design field, there are many multipliers available to increase the performance level. In this chapter, approximate compressors used in a parallel multiplier are proposed. The two new approximate 4-2 compressors propose that the simplified compressors have better power consumption than the optimized 4-2 compressor existing designs. The renovation module of a parallel multipliers are going to use these approximate compressors. For a parallel multiplier, four different outlines exploiting the proposed approximate compressors are proposed and analyzed. The design of multiplier relies on the compressor and can meet with the respect to the circuit-based design. The results of the proposed design show and accomplish substantial declines in delay, transistor count, and power dissipation compared to an exact compressor design; likewise, two of the future multiplier designs offer an brilliant proficiencies for multiplication of an image.

19.1 INTRODUCTION

Digital logic circuits are used in the implementation of most computer arithmetic applications, thus operating with a good precision and high degree of reliability. However, many applications such as image processing and multimedia can tolerate errors and imprecision in computation and still produce useful and meaningful results. Algorithms and accurate and precise models are not always efficient or suitable for use in these applications. The model of approximate computation relies on relaxing completely deterministic building modules and fully precise when, for example, designing energy efficient systems. This allows imprecise computation to redirect the existing design process of systems and digital circuits by taking advantage of a decrease in cost and complexity with possibly a potential increase in power efficiency and performance.

In today's digital signal processing and various other applications, multipliers play an important role. With the advances in technology, many researchers are trying to design multipliers which offer either of the following design targets—low power consumption, high speed, regularity of layout, and hence less area—or even combination of them in single multiplier which makes them suitable or efficient for various low power, high speed, and compact very large-scale integration implementation. In computer arithmetic operations, multiplication and addition are widely used; for approximate computing, full adder cells have been extensively analyzed for addition.[1,2–4] These adders compared and proposed several new metrics for evaluating probabilistic and approximate adders with respect to the unified figures of merit for design assessment for approximate computing applications. For each input to a circuit, the error distance (ED) is defined as the arithmetic distance between the correct output and an erroneous output.[1] The trade-off between power and precision has also been quantitatively evaluated in Ref. [1].

However, the design of inexact multipliers has received less attention. Multiplication is the repeated sum of partial products; however, the straightforward application of inexact adders when designing an approximate multiplier is not good because it would be very inefficient in terms of hardware complexity, precision, and other performance metrics. Several approximate or inexact multipliers are proposed in the literature.[4–7] A truncated multiplication method uses most of these design models; they estimate that the least significant columns of the partial products are constants. In Ref. [4], neural network application uses an imprecise array multiplier by omitting some of the least significant bits in the partial products (in the array, some adders are removed). In Ref. [5], a truncated multiplier is proposed with a correction constant. This design computes the sum of the $n + k$ most significant columns of the partial products and truncates the other $n - k$ columns for an $n \times n$ multiplier. Then, the result of $n + k$ bit is rounded to n bits. In the next step, the rounding error (i.e., the error is generated by rounding the result to the n bits) and reduction error (i.e., the error is generated by truncating the $n - k$ least significant bits) are found. To the estimated value of the sum of these errors to reduce the ED, the correction constant ($n + k$ bits) is selected to be as close as possible.

If the partial products in the $n - k$ least significant columns are all ones or all zeros, then the truncated multiplier with constant correction has the maximum error. In Ref. [6], a variable-correction-truncated multiplier has been proposed, which changes the correction term based on the column $n - k - 1$. If all partial products in column $n - k - 1$ are one, then the correction term gets increased. Similarly, if all partial products in this column are zero, the correction term gets decreased. In Ref. [7], a simplified (and thus approximate) multiplier block is proposed for building a larger multiplier arrays. Compressors have been widely used in the design of a fast multiplier[8] to speed up the partial product reduction tree and to reduce the power dissipation. In Refs. [8] the optimized designs of 4-2 exact compressors have been proposed. Refs. [7, 8] are also considered as compression for approximate (inexact) multiplication. An approximate signed multiplier has been proposed in Ref. [7], for use in an arithmetic data value speculation; the Baugh–Wooley algorithm is used for multiplication process. However, for the compressors, no new design is proposed for the inexact computation. In Ref. [8], designs of approximate compressors have been proposed, which do not target multiplication. It should be noted that the approach of Ref. [7] improves over Refs. [7, 8] by utilizing a simplified multiplier block that is amenable to inexact or approximate multiplication.

19.2   COMPRESSOR DESIGN

19.2.1   EXACT COMPRESSOR

As shown in Figure 19.1, the exact 4-2 compressor[9] has five inputs A, B, C, D, and C_{in} to generate three outputs sum, carry, and C_{out}. The four inputs A, B, C, and D and the output sum have similar weight. The input C_{in} is the output from the preceding lower widespread compressor and the C_{out} propagates for the compressor inside the next significant level.

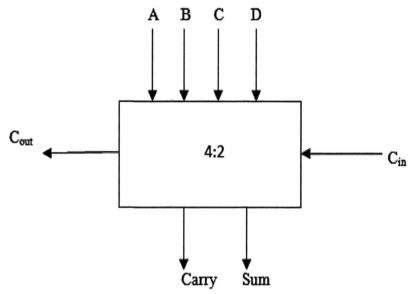

FIGURE 19.1   4-2 Compresssor.

19.2.2   APPROXIMATE COMPRESSOR DESIGN 1

In design 1, we make carry' $= C_{in}$ by changing the values of output with this approximation that the output carry in an exact 4-2 compressor has the same value as of input C_{in}. We can reduce the complexity of the design as well as the difference between exact and approximate outputs by making sum value to 0 (Fig. 19.2).[3]

Dadda multiplier using design 1: An unsigned 8 × 8 Dadda tree multiplier is revised to approach the impact of using the proposed compressor for an approximate multipliers. Initially, the proposed multipliers are used to generate all partial products, and gates are used. The 4-2 compressors, full adders, and half adders are used by the reduction part; each partial product bit is denoted by a dot. In the first stage, eight compressors, two full adders, and two half adders are used to decrease the scale of partial products into no more than four rows.

In the second or last stage, 10 compressors, 1 full adder, and 1 half adder are used to figure the two final rows of the partial products. Therefore, 2 stages of reduction and 3 half adders, 3 full adders, and 18 compressors are required in an 8 × 8 Dadda multiplier for the reduction circuitry (Fig. 19.3).

$$\text{Carry}' = \overline{\overline{x1x2} + \overline{x3x4}}$$

$$\text{Sum}' = \text{Cin}(\overline{\overline{x1 \oplus x2} + \overline{x3 \oplus x4}})$$

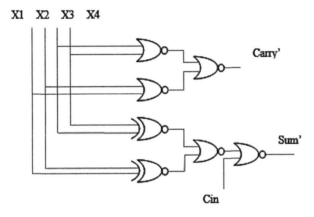

FIGURE 19.2   Gate level design 1 compressor.

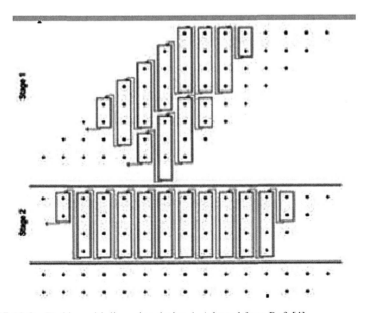

FIGURE 19.3   Dadda multiplier using design 1. Adapted from Ref. [4].

19.2.3   APPROXIMATE COMPRESSOR DESIGN 2

In the proposed design, we approximated carry' and C_{in} as they have the same weight; here, we take C_{in} as 0 so that we can remove the carry', hence performance increased by reducing the error rate.

$$\overline{Sum} = Cin(\overline{x1 \oplus x2} + \overline{x3 \oplus x4})$$

$$Carry' = \overline{x1x2 + x3x4}$$

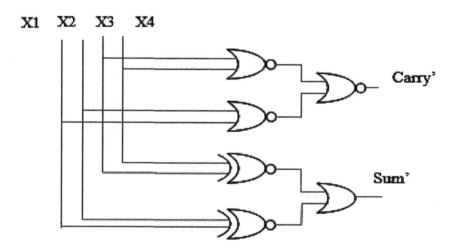

FIGURE 19.4   Approximate compressor design 2.

In design 2 (Fig. 19.4),[3] it has three cases: In the first case of computation (multiplier 1), design 1 is utilized for all 4-2 compressors. In the second case of computation (multiplier 2), design 2 is used for the 4-2 compressors; since C_{in} and C_{out} are not considered in the design 2, a low number of compressors are used in this multiplier to reduce circuit complexity; in this design, 17 compressors, 1 full adder, and 6 half adders are used. In the third case of computation (multiplier 3), design 1 is used for the compressors in the $n - 1$ least significant columns. The exact 4-2 compressors are used by the other n most significant columns in the reduction circuit.

19.3 PROPOSED DESIGN

In this proposed implementation, a new carry save adders are used for the partial products and then hybrid design in that internal design 1 and design 2 compressors is used likewise in truncated multiplier. Through that, it will get more efficiency and high performance. The future scope is that instead of 4-2 compressor, there are 5-2 compressor and the main applications like image processing and compression (Fig. 19.5).[4]

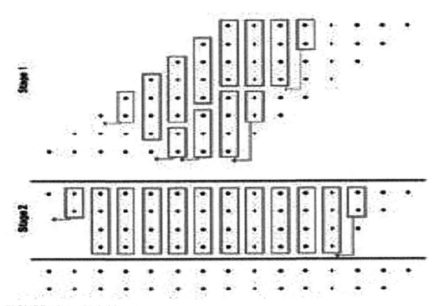

FIGURE 19.5 Hybrid compressor.

19.4 SIMULATION RESULTS

These circuits are designed and performed by using ModelSim software and synthesized by using Xilinx software. Simulation results of exact compressor are shown in Figure 19.6. For inputs $A = 1$, $B = 0$, $C = 1$, $D = 0$, and $C_{in} = 1$, the outputs are carry = 0, sum = 1, and $C_{out} = 1$. C_{out} propagates from one stage to next stage.

Simulation results of approximate compressor (design 1) are shown in Figure 19.7. For inputs $A = 1$, $B = 0$, $C = 0$, $D = 1$, and $C_{in} = 1$, the outputs are carry$_B = 1$, sum$_B = 0$, and $C_{out,B} = 1$. $C_{out,B}$ propagates from one stage to next stage.

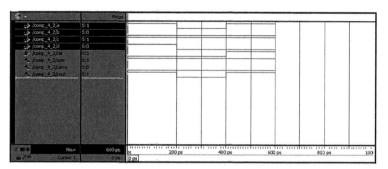

FIGURE 19.6   Simulation results of exact compressor.

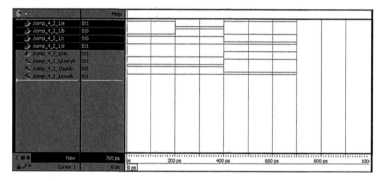

FIGURE 19.7   Simulation results of approximate compressor (design 1).

Simulation results of approximate compressor (design 2) are shown in Figure 19.8. For inputs $A = 1$, $B = 1$, $C = 1$, and $D = 0$, the outputs are carry$_B$ = 1 and sum$_B$ = 0. In design 2, C_{in} and C_{out} are considered to have same weights, so $C_{in} = C_{out} = 0$ (Table 19.1).[3]

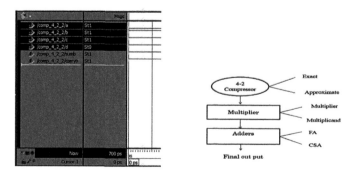

FIGURE 19.8   Simulation results of approximate compressor (design 2).

Simulation results of multiplier 1 are shown in Figure 19.9. Here, *A*, *B*, and clock are the inputs and *y* is the output.

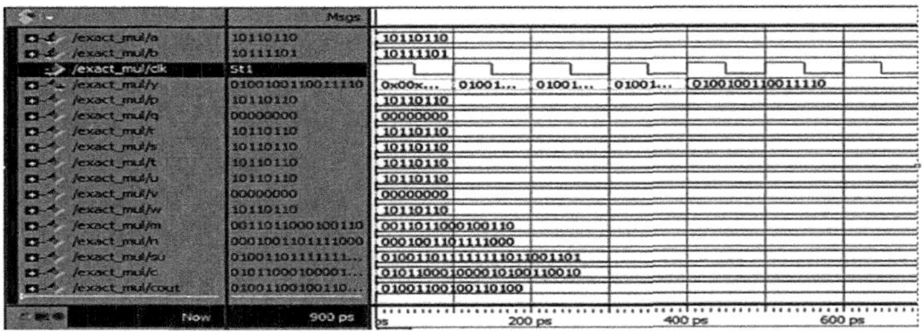

FIGURE 19.9   Simulation results of multiplier 1.

Simulation results of multiplier 2 are shown in Figure 19.10.

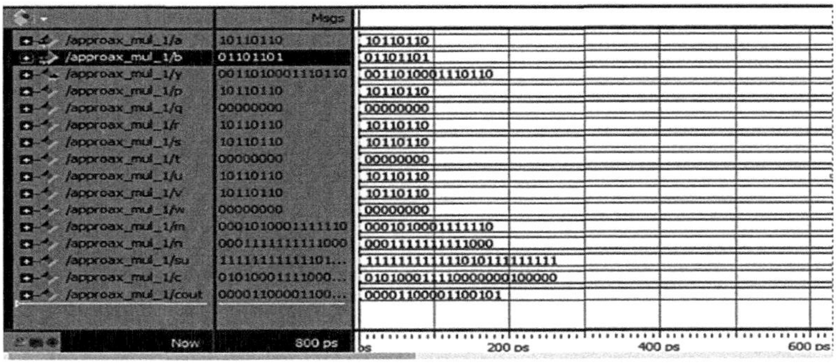

FIGURE 19.10   Simulation results of multiplier 2.

Simulation results of multiplier 3 are shown in Figure 19.11.

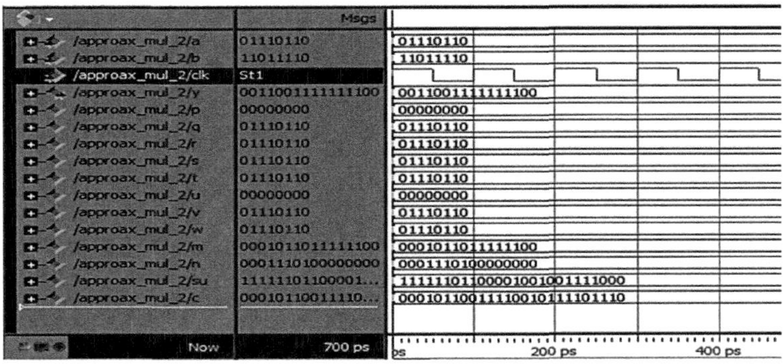

FIGURE 19.11   Simulation results of multiplier 3.

Simulation results of multiplier 4 are shown in Figure 19.12.

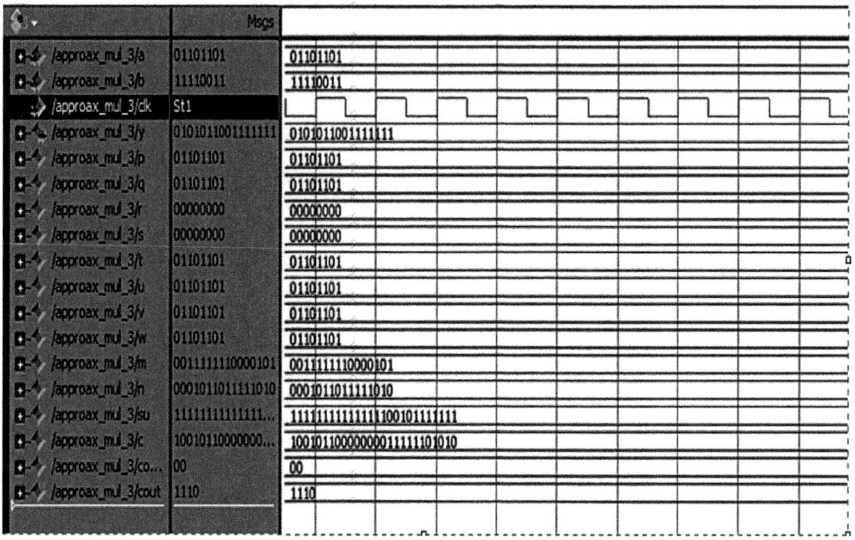

FIGURE 19.12   Simulation results of multiplier 4.

TABLE 19.1   Comparison of Delay in Compressor Designs. Adapted from Ref. [4].

Design of compressor	Delay (ns)
Approximate compressor design 1	30.07 ns
Approximate compressor design 2	29.173 ns
Approximate compressor design 3	28.778 ns

19.5   CONCLUSION

In this chapter, we are using an approximate (inexact) compressor. By using design 1 and design 2, four 8×8 bit approximate or inexact multipliers are designed to reduce the circuit complexity and to increase the performance. Design 3 multiplier was implemented as a proposed method. ModelSim is used to generate the simulation results. Inexact or approximate computing is a prominent model for computation at nanometer technology. A significant operational advantage is offered by computer arithmetic for inexact computing; an extensive literature exists on inexact or approximate adders.

These inexact compressors are used in the reduction module of four approximate multiplier, and approximate compressors show a reasonable reduction in power consumption, delay, and transistor count compared to an exact design. Furthermore, the application of these approximate multipliers is image processing.

KEYWORDS

- **exact compressor**
- **parallel multiplier**
- **approximate compressor designs**
- **DSPs**
- **Baugh–Wooley algorithm**

REFERENCES

1. Kyaw, K. Y.; Goh, W. L.; Yeo, K. S. Low-power High-speed Multiplier for Error-tolerant Application. In *IEEE International Conference of Electron Devices and Solid State Circuits (EDSSC)*, 2010.
2. Lau, M. S. K.; Ling, K. V.; Chu, Y. C. Energy-Aware Probabilistic Multiplier: Design and Analysis. In *Proceedings of the 2009 International Conference on Compilers, Architecture, and Synthesis for Embedded Systems*, Grenoble, France, 2009; pp 281–290.
3. Kulkarni, P.; Gupta, P.; Ercegovac, M. D. Trading Accuracy for Power in a Multiplier Architecture. In *24th International Conference on VLSI design*, 2011.
4. Ma, J.; Man, K.; Krilavicius, T.; Guan, S.; Jeong, T. Implementation of High Performance Multipliers Based on Approximate Compressor Design. In *Presented at the Int. Conf. Electrical and Control Technologies*, Kaunas, Lithuania, 2011.
5. Parhami, B. *Computer Arithmetic Algorithms and Hardware Designs*, 2nd ed. Oxford Univ. Press: London, 2010.
6. Liang, J.; Han, J.; Lombardi, F. New Metrics for the Reliability of Approximate and Probabilistic Adders. *IEEE Trans. Comput.* **2013,** 63 (9), 1760–1771.
7. Chang, C.; Gu, J.; Zhang, M. Ultra Low-Voltage Low-Power CMOS 4-2 and 4-5—Compressors for Fast Arithmetic Circuits. *IEEE Trans. Circuits Syst.* **2004,** 51 (10), 1985–1997.
8. Radhakrishnan, D.; Preethy, A. P. Low-Power CMOS Pass Logic 4-2 Compressor for High-speed Multiplication. *Proc. 43rd IEEE Mid 1st Symp. Circuits Syst.* **2000,** 3, 1296–1298.
9. Prasad, K.; Parhi, K. K. Low-power 4-2 and 5-2 Compressors. In *Proc. 35th Asilomar Conf. Signals, Syst. Comput.* **2001,** 1, 129–133.

CHAPTER 20

COST-EFFECTIVE IMPLEMENTATION OF DIGITAL KARAOKE

P. SANDEEP*, B. SAI CHAKRADHAR, and MD. SHARUQUE

Department of ECE VITS, Hyderabad, Telangana, India

Corresponding author. E-mail: san9ap@gmail.com

ABSTRACT

This paper presents an implementation of low-cost digital karaoke machine for vocal, which is capable of removing the voice component of a music file and storing the user's singing voice with the background music to an external compact flash memory. It can also be used as a stand-alone voice recorder where the playback sound obtained consists of the original music and recorded sound. The process being vocal removed from the music file from source and the remaining background music is added to the recorded sound signal acquired from the microphone. Relatively high recording speed around 20 KHz is required in order to achieve a high-quality sound. An external serial multichannel ADC is used to extract the music signal. Storing is performed using an external storing device, like a memory flash card; flash memory by considering the size of data as local microcontroller memory does not have enough space to store such a big file.

20.1 INTRODUCTION

Karaoke is a form of entertainment in which users sing along with recorded music using a microphone. The basic idea of implementation is to create some tape of music without vocals for an event and to achieve this removal of vocal from records, the karaoke machine is helpful. During stereo recording, the singer is usually placed in the middle of the left and right microphones where the voice is simultaneously and equally fed to both the channels.

Hence, the vocal component is almost identical to both the left and right tracks of the audio song. While the majority of the instrumental sounds are slightly different in two channels,[1] required output may not be applicable. So by subtracting the left and right channel, the vocal will be removed and the music shall remain serving the basic purpose of the machine.

The technique used may not be perfectly preferable in all the cases based on the assumption about the voice of the singer resulting in the echoes. These echoes leading to unwanted noisy sound remain in the background which makes difficulty in removal of vocals.

In this chapter, removal of the vocal is done in two phases. The first phase of the karaoke machine involves removing the original voice of the artist in a song and the second phase is the implementation by using a PIC16F73 microcontroller.

20.2 SYSTEM CONFIGURATION

The execution of this method is performed using several hardware components and process of removing the vocal.[2] The two phases of obtaining the required output can be achieved using the proposed hardware.

20.2.1 HARDWARE REQUIREMENTS

The basic components required for the implementation of two phases include a microcontroller, amplifier, and converters. The block diagram is as shown in Figure 20.1.

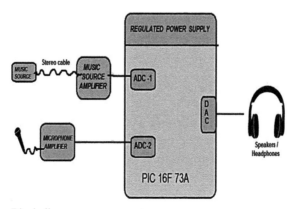

FIGURE 20.1 Block diagram.

20.2.1.1 MICROCONTROLLER

PIC16F73A is an FLASH-based 8-bit microcontroller which is a very efficient chip of 28 pins consuming +5 V of the power supply and speed of 20 MHz. Pin diagram of PIC16F3[3] is shown in Figure 20.2.

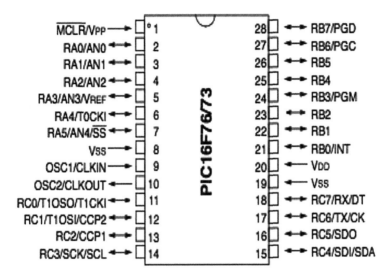

FIGURE 20.2 Pin diagram.

It consists of the following configurations:

- 2 PWM 8 bit
- 256 bytes EEPROM data memory
- 25 mA sink/source per I/O
- In circuit debug
- Self-programming
- Parallel slave port[4]

It serves the logical functions as follows:

- Analog-to-digital conversions (ADCs) as the controller unit consists of two internal ADC
- Synchronization using 11.05 MHz crystal oscillator
- PWM technique is been performed

- RESET technique is also performed and by using programming code
- Digital-to-analog converter (DAC) of R-2R ladder is also internally connected to the microcontroller unit

The microcontroller takes the control operation to the amplifier section, microphone, and speakers.

20.2.1.2   AMPLIFIER LM386

- Features of the amplifier are as follows:
- Battery operation
- Minimum external parts
- Wide supply voltage range: 4–12 or 5–18 V
- Low quiescent current drain: 4 mA
- Voltage gains from 20 to 200 ground referenced input
- Self-centering output quiescent voltage
- Low distortion: 0.2% (AV = 20, VS = 6 V, RL = 8 Ω, PO = 125 mW, f = 1 kHz)
- Available in eight pins MSOP package

The amplifiers can be music source amplifier taking input from music source and giving to the microcontroller through ADC. It can also be microphone amplifier from the microphone to ADC of PIC.[5]

20.2.2   SOFTWARE TOOLS

The software tools required to execute the process are as follows:

- PIC C Compiler
- Simulink
- PIC kit 2—Programmer/Burner
- Express Schematic (SCH)—SCH drawing tool

The compiler is used for writing the appropriate code in executing the process including C language. Simulink to view the formats of the original song and karaoke song obtained. The programmer/burner is used to burn the program onto the microcontroller kit.[6]

20.2.3 IMPLEMENTATION

The process involved in karaoke machine is given in Figure 20.3.

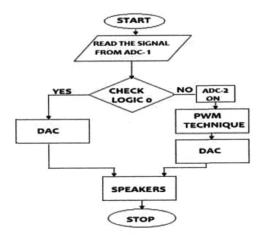

FIGURE 20.3   Flow chart of karaoke machine.

The process performed in the execution starts with the input for the machine provided from the source like mobile phone or iPad or CD player, through a stereo jack, and connected to the microchip by segregation into three various channels:

- Data
- Power supply
- Ground

This music file is then passed through the amplifier of LM386 circuit that increases the strength of the signal. As the signal exists in only positive cycles, it is fed to ADC-1 of PIC that converts the analog form into digital.

The other input is from a microphone connected through microphone amplifier of LM386 circuit, which raises the strength of the voice signal. This is then carried forward to ADC-2 that converts analog-to-digital format because the voice signal should exist in positive cycles.[7] The ADCs connected are used to synchronize both outputs and been reversed into one DAC.

The DAC is connected to a speaker/headset to listen to the output of the machine so that clarity exists more in speakers of 8 Ω.

20.3   RESULTS

The hardware kit for the process is as shown in Figure 20.4.

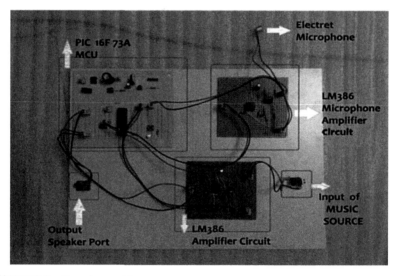

FIGURE 20.4   Karaoke machine.

Data taken from the input, modifying the signal and getting a karaoke signal at the output, are done by executing in Simulink. The output window shown in Figure 20.5 gives the input original song and output karaoke song.

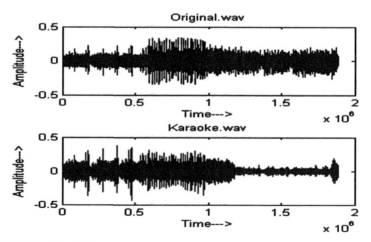

FIGURE 20.5   Simulink output.

The output karaoke song is mainly intended to separate speech and music from the original song and add the user's voice.

20.4 CONCLUSION

This proposed chapter presents the simple design of a cost-effective model of karaoke machine capable of removing the voice component of a music file and adding a desirable singing voice with the background music. This system has several applications in the field of entertainment, film-editing applications, music industry, or several industrial applications which promote cheaper entertainment products, fast response, and less manual power. It also reduces the complexity in arranging live concerts by improving the efficiency and clarity. The proposed work can be implemented by application-specific integrated circuit so a dedicated hardware can implement at a low cost.

KEYWORDS

- **karaoke**
- **ADC**
- **flash memory**
- **microphone**
- **memory**
- **microcontroller**

REFERENCES

1. Bhalani, N. R.; Singh, J.; Tiwari, M. Implementation of Karaoke Machine on the DSK6713 Processor. *Int. J. Comput. Appl.* (0975–8887).
2. www.ti.com/lit/ds/symlink/lm386.pdf.
3. Thenua, R. K.; Agrawal, S. K.; Member. Hardware Implementation of Adaptive Algorithms for Noise Cancellation. In *IACSIT*, vol. 2, no. 2, March 2012.
4. Jadiye, S.; Bakshi, A.; Gogate, M. Karaoke Machine Implementation and Validation Using Out of Phase Stereo Method along with LMS. In *2014 IJEDR*, vol. 3, no. 1, 2014, ISSN: 2321–9939.
5. http://www.microchip.com/wwwproducts/en/PIC16F73.

6. Shenoy, A. In *Singing Voice Detection for Karaoke Application,* the Proceedings SPIE 5960,Visual Communications and Image Processing, Bellingham, WA, USA, 2005,.
7. Jadiye, S. Vocal Removed from Multi Object Audio. *Int. J. Rec. Trends Eng. Res.* **2016,** *2* (6), 424–428 [ISSN: 2455–1457].

CHAPTER 21

DEVELOPING A SIMPLE AND ECONOMIC VOICE CONTROL MECHANISM FOR OPERATING HOME APPLIANCES

D. DURGA BHAVANI*

Department of Computer Science and Engineering, CVR College of Engineering (Autonomous), Hyderabad, Telangana, India

Corresponding author. E-mail: drddurgabhavani@gmail.com

ABSTRACT

Smart homes/home automation is an emerging market already resulting in substantial growth in users and customers. The need for smart homes in any public is always high. In particular, home automation is regarded as a desirable solution to keep home safe especially while they are not occupied—either using the day when the inhabitants are at work or for extended periods when they are on vacation. Smart surveillance features also allow the safety of children and elderly people to be monitored remotely via smartphone applications. This paper will give you a way to showcase your home appliances in a smarter and in an effective way. It reduces the unnecessary energy consumption and ensures that the appliances are under control. All it needs is a microcontroller, voice command through your android application in your mobile, the commands will be processed, and corresponding action will be triggered. It is cost effective as well. When compared, the other home automation appliances are very expensive. It eases a user's interaction with home equipment by providing a new, easy to use interface. The initial phase of the development involves surveying the current field by conducting speech-recognition software and the needs of potential users. These studies provided a foundation for the development of the idea.

21.1 INTRODUCTION

Voice recognition takes input as the spoken word to a computer program. This procedure is essential to virtual reality since it gives a genuinely common and instinctive method for controlling the simulation while enabling the user's hands to stay free. This chapter will explore the uses of voice recognition in the virtual reality field, observe how voice recognition is accomplished, and list the academic disciplines that are central to the indulgent and development of voice recognition expertise.

Voice activation for home appliances has only just started with electronics companies like Samsung releasing a voice-controlled TV remote. Taking a step further into home automation, using voice activation emancipates the innovation in IOT revolution (*Speech Recognition Setup—Microsoft*).[1]

In particular, home automation is regarded as a desirable solution to keep home safe while they are not occupied—either using the day when the inhabitants are at work or for extended periods when they are vacationing. Smart surveillance features also allow the safety of children and elderly people to be monitored remotely via smartphone apps. In integrated townships, the smart security features must be integrated with larger security measures at the project level.—Managing Director, Pride Group, *The Hindu* (*Article No.: 8686620*. http://www.thehindu.com/features/homes).

This chapter gives you an efficient and an effortless way of controlling your appliances. The basic parts include the voice recognition. All that is required keeping in mind is that the end goal to utilize discourse acknowledgment is an advanced mobile phone with some type of discourse acknowledgment application. Most telephones even accompanied a discourse acknowledgment application pre-introduced. With voice-controlled home mechanization, everybody can lead a more agreeable life. Circling the house to turn on every one of the lights can be tedious; it would rather be less demanding to utilize your voice. This chapter is especially valuable for individuals with handicaps because the advantage may be considerably more noteworthy since they would now be able to do things that they, because of their inability, couldn't do in the recent past. This sort of voice-controlled framework could enhance their personal satisfaction, while diminishing the need for help is a well-known and reachable thought to most people in western culture.

All that is required keeping in mind is that the end goal to utilize voice recognition is an advanced smart mobile phone. Most smart phones even accompanied with a built-in voice recognition application. Human being can lead a comfortable life with voice-controlled home automation. To turn on/off the lights, humans roam around the house, which can be a very tedious job. Instead of that, we use voice. This sort of voice-controlled framework could enhance their personal satisfaction, while moving back the need for help.[2,3]

21.2   RELATED WORK

A detailed study and comparisons on the available solutions have been made. Some of the existing home automation devices are discussed below.

1.   Apple Homekit: Works on Siri—iPhone controlled

Homekit is a bunch of devices manufactured for supporting iPhone features. All are voice controlled based. To control the devices using the Homekit, there is a need of an Apple TV to act as a bridge. These smart kit facility is only limited to Apple users. This results in cost overflow. These are mostly preferred in foreign countries.

2.   Athom Homey: Virtual assistant like Cortana—$243

The voice interface is Power BI technology and machine-learning Cortana connects all devices, including Xbox One and PC. This is a product from Apple.

3.   Insteon: Provides a network of configurable devices

Insteon Hub is a trending smart home automation series which monitors the home with more convince. To initiate an Insteon Hub, all the bulbs, plug-in, and all the appliances should be replaced with the Insteon sensors which results in much complication.

The following are the problems faced by the users in using the above smart home devices:

- The existing products are not cost-effective and the need of Internet is necessary.
- To change a home to a smart one, the whole wiring has to be changed which increases the complexity.
- Some of the existing applications are very intricate.

21.2.1   SURVEY CONDUCTED BY NCAER

A research was done on smart homes in India by National Council for Applied Economics Research (NCAER) 2015, which highlighted the following points:

- GDP growth in middle-class economy sector for FY1: 7.6%
- Urban middle-class population: 23.6 million
- Middle-class households: 113.8 million, 547 million.
- Indians constitute 3% of global middle-class sector with annual per capita: Rs. 61,480
- Growth of smart homes in Tier 1 cities: 15–18%
- Growth of smart homes in Tier 2 and 3 cities: 5–10%

21.2.2   SURVEY CONDUCTED BY THE HINDU

A survey has been conducted by *The Hindu* in collaboration with Center for Good Governance (CGG) which has stated that average Indian home buyers are positive to the idea of a smart home; they are willing to spend no more than 1–3% extra for home featuring smart solutions whereas upper income groups are willing to spend up to 5–8%. The survey was conducted across 21 states and the lower margin was set for 2$/day and higher margin was 13$/day and a monthly income of 2–10 lakhs.

Prediction of *The Times Magazine*: US$ 12 billion in the upcoming 5 years.

A survey in Bengaluru of 220 projects offering a home costing more than crores of rupees last year revealed that 80 of them were constructing fully automated homes. According to Madhav Rao, National Secretary of Indian Society of Heating Refrigeration & Air Conditioning Engineers (ISHRAE), construction of smart homes is projected to have a growth rate of 35–40%. By 2016, 300 projects will have fully automated homes. ISHRAE organized 2018 Acrex India in Bengaluru between February 22 and 24 enabling 600 exhibitors to showcase the components going into smart homes which clearly shows the impact on smart homes (http://www.cnet.com/news/talk-to-your-house-with-these-voice-activated-smart-home-systems).

21.3   VOICE RECOGNITION

Voice recognition is "the technology by which sounds, words or phrases spoken by humans are converted into electrical signals, and these signals are transformed into coding patterns to which meaning has been assigned." Here, human voice is used to communicate with others in their immediate surroundings. Voice recognition uses vocal interaction that is already present. The proposed solution uses an Arduino microcontroller and a Bluetooth module which is more cost-effective. This product just needs

- an app to give command
- Bluetooth to connect
- Arduino to trigger

21.3.1 PROBLEM BEING SOLVED

- This chapter shows us a smart way of showcasing our appliances. It makes a path to make our home a smart one with a zero percent complexity. Other smart devices once they face any problem the only thing that can be done is replacing the device which increases the cost.
- It reduces the common problem of the switch boards by ditching the conventional haystack of electrical wires. To turn our home into a smart one, all the wiring has to be replaced. The major drawback is making the customers think twice.
- This chapter is eliminating the need for smart homes in the middle-class public of the state. Previous smart devices experienced a bad impact in the area of cost. This is considered as the major drawback. This problem is solved in this chapter.

21.3.2 ARCHITECTURE

The proposed technology includes the following:

1. *Arduino Uno microcontroller with ATmega 328 processor*

Arduino/Genuino Uno is a microcontroller board based on the ATmega328P. It has 14 digital input/output pins (of which 6 can be used as pulse width modulation outputs), 6 analog inputs, a 16-MHz quartz crystal, a USB connection, a power jack, an in-circuit serial programming header, and a reset button. It contains everything needed to support the microcontroller; simply connect it to a computer with a USB cable or power it with an AC-to-DC adapter or battery to get started (*Arduino for Dummies by John Nussey*; *Arduino uno*, https://www.arduino.cc/en/Main/Software).

2. HC-05 Bluetooth module

Bluetooth is a wireless technology standard for exchanging data over short distances (using short-wavelength ultrahigh frequency radio waves in the industrial, scientific, and medical band from 2.4 to 2.485 GHz) from fixed and mobile devices and building personal area networks. Range is approximately 10 m (30 ft). These low-cost Bluetooth submodules work well with Arduino and other microcomputers (Bluetooth module, https://www.olimex.com/Products/Components/RF/BLUETOOTH-SERIAL-HC-06/resources/hc06.pdf).

3. 12-V 4-channel relay board for controlling the switches (*Handbook for Digital IC's from Analogic Devices*)

4. Android voice recognition, speech-to-text (STT)–text-to-speech TTS technology

In this chapter, voice commands play a major role in controlling the appliances. These voice commands are sent through our smart phones. Once the commands are received by the Arduino, it triggers the command and the appliances respond to it (Fig. 21.1).[1]

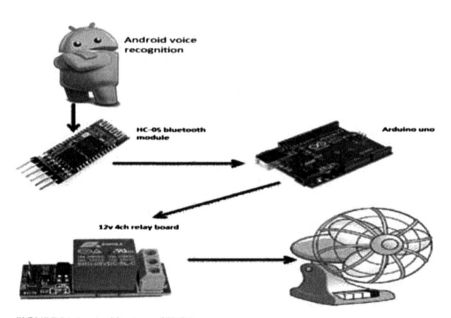

FIGURE 21.1   Architecture of VoCA.

First thing to be done is connecting with the Bluetooth. The flow starts with giving the command through your android application, then the command is triggered by the Arduino, and then the appliance connected to the relay board responds accordingly (*Android Studio: How to Guide and Tutorial by Clive Sargeant*; *ATMEL 328P Data sheets*).

21.4   KEY FEATURES

21.4.1   CHOICE OF TECHNOLOGY

The technology used Arduino and Bluetooth. The current solution is using high-level integrated chips.

21.4.2   ON SIZE

Size of my device is the size of a grown man's palm, the existing solutions are bulkier. This adds to the effortless use of voice output communication aid (VoCA).

21.4.3   ON COST

All of the current solutions cost more than $240. My product provides excellent cost-effectiveness.

21.5   CONCLUSION AND FUTURE WORK

Such credible portability and minute for currently available solutions for home automation are not available. Indian home automation companies like Home Automat and Z-Wave do not provide such simple and cost-effective solutions. Promoting the product on social media platforms and conducting awareness for simple and cheap solutions and a live comparison for smart homes for my target customers will get my product sold.[1] This work deals with one language and a microcontroller which has 342 kb of memory. This can be further extended using Wi-Fi connection that benefits more to commercial institutions and organizations. This chapter can be more useful by making this work for all the languages and also for the accents which make people more comfortable using it. Any product having the above features that eliminates the need of smart homes within the budget would definitely be considered as the best one.

KEYWORDS

- **smart home**
- **home automation**
- **Internet of Things**
- **voice recognition**
- **Arduino Uno microcontroller**
- **android app development**

REFERENCES

1. Ankitha, A. M. Voice Controlled Home Appliances. *Int. J. Electr. Electron. Comput. Sci. Eng.* **2014,** *1* (1), 32–34.
2. Fourcin, A.; Harland, G.; Barry, W.; Hazan, V., Eds. *Speech Input and Output Assessment*; Ellis Horwood Limited: Chichester, UK, 1989.
3. Yannakoudakis, E. J.; Hutton, P. J. *Speech Synthesis and Recognition Systems*; Ellis Horwood Limited: Chichester, UK, 1987.

CHAPTER 22

MULTILEVEL BOOST CONVERTER IMPLEMENTATION FOR PHOTOVOLTAIC APPLICATIONS

BHARATHA SATEESH[1*] and PRABHU G. BENAKOP[2]

[1]Department of ECE, Vaagdevi College of Engineering, JNTU, Warangal, Telangana, India

[2]Department of ECE, Indur Institute of Engineering and Technology, Ponnala (V), Siddipet, Medak, Telangana, India

[*]Corresponding author. E-mail: basateesh27@gmail.com

ABSTRACT

This chapter presents the implementation of a maximum power point tracker (MPPT) for photovoltaic (PV) applications by using multilevel boost converter and FPGA board. The control algorithm for extracting maximum power from the cell is proposed by means of the very high speed integrated hardware description language code and implemented using Xilinx XC3S400 FPGA board. In this work, a practical implementation of the real-time estimate, perturb, and absorb algorithm for maximum power point tracking control in a PV system has been developed. The developed implementation has the advantage of simple programming with high performance even with low-resolution analog to digital converter and low-cost current sensor. The proposed technique has been validated through detailed experimental work.

22.1 INTRODUCTION

Renewable energy sources such as solar energy are acquiring more significance due to shortage and environmental impacts of conventional fuels. The photovoltaic (PV) system for converting solar energy into electricity is in general costly and is a vital way of electricity generation only if it can produce the maximum possible output for all weather conditions. The PV array has a highly nonlinear current–voltage characteristic varying with the irradiance and temperature that substantially affects the array power output. The maximum power point tracking (MPPT) control of the PV system is therefore critical for the success of a PV system. Perturb and observe (P&O) MPPT algorithm which is implemented by XILINX FPGA has been considered extensively in the literature. Field programmable gate arrays (FPGAs) are standard integrated circuits that can be programmed by a user to perform a variety of complex logic functions. The high level of integration available with these devices (currently up to 500,000 gates) means that they can be used to implement complex electronic system. Furthermore, there are many advantages due to the rapid design process and reprogrammable function. XILINX FPGA enables to produce prototype logic designs right in a short period. It is possible to create, implement, and verify a new design. A configuration program stored in internal static memory cells of the XILINX FPGA is written by very high speed integrated hardware description language (VHDL) programming language that has been designed and optimized for describing the behavior of digital systems; VHDL has many features appropriate for describing the behavior of electronic components ranging from simple logic gates to complete microprocessors and custom chips. Features of VHDL allow electrical aspects of circuit behavior (such as rise and fall times of signals, delays through gates, and functional operation) to be precisely described. The utilization of FPGAs instead of other architectures was mainly based on four factors: the acceleration of the design or parts of it, the flexibility of reconfiguration hardware, the reduction of costs, and the energy consumption. These factors had a different impact on each application area.[1] Several MPPT techniques have been proposed, during the last decades. They range from conventional methods, from simple hill-climbing algorithms (P&O, MP&O, and estimate–perturb–perturb), to fuzzy logic and neural network algorithms. The hill-climbing algorithm[2–4] is widely used in practical PV systems due to its simplicity and as it does not require prior study or modeling of the source characteristics and can account for characteristics drift resulting from aging, shadowing, or other operating irregularities, but its performance is poor compared to artificial intelligent methods, Recently, the increasing performance and cost reduction of digital circuits have made possible their applications for power converter control. Comparing with other digital signal processors, FPGA-based systems could provide a number of run-time advantages over the sequential machines such as a microcontroller. Moreover, with concurrent operation, it is executed continuously and simultaneously which is faster than DSP. Thus, the FPGA has been applied for high-speed switching circuit to reduce equipment sizing,[5] especially in the implementation of maximum power point (MPP), FPGA features are utilized.[6,7] The chapter is organized as follows. Section 22.2 shows the system configuration of the proposed system. In Section 22.3, a multilevel boost converter (MLBC) is analyzed. The MPPT control is discussed in Section 22.4. Section 22.5 shows the experimental results. Finally, conclusions are presented in Section 22.6.

22.2   PV CELL MODEL AND SIMULATION

The simplest equivalent circuit of a solar cell is a current source in parallel with a diode. The output of the current source is directly proportional to the light falling on the cell. The diode determines the characteristics of the cell. Increasing sophistication, accuracy, and complexity can be introduced to the model by adding in turn:

- temperature dependence of the diode saturation current I_0
- temperature dependence of the photo current I_L
- series resistance R_s, which gives a more accurate shape between the maximum power point and the open circuit voltage
- shunt resistance R_p in parallel with the diode
- Either allowing the diode quality factor n to become a variable parameter (instead of being fixed at either 1 or 2) or introducing two parallel diodes (one with $A = 1$, one with $A = 2$) with independently set saturation currents.[8] For this research work, a model of moderate complexity was used. The model includes temperature dependence of the photo current I_L and the saturation current of the diode I_0. A series resistance R_s was included, but not a shunt resistance. A single shunt diode was used with the diode quality factor set to achieve the best curve match. The circuit diagram for the solar cell is shown in Figure 22.1.

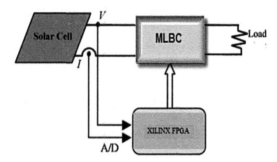

FIGURE 22.1   Configuration of the controlled MPPT.

22.3   MLBC DESIGN

Figure 22.2 illustrates a MLBC which combines the boost converter and
the switched capacitor function to provide an output of several capacitors
in series with the same voltage and self-balanced voltage. The major advantages of this topology are (1) continuous input current and (2) a large conversion ratio with low duty cycle and without a transformer. It can be built in
a modular way and more levels can be added without changing the main
circuit; it provides several self-balanced voltage levels and only one switch
is necessary. The MLBChas a higher conversion ratio with the conventional
converter based on the number of level used.

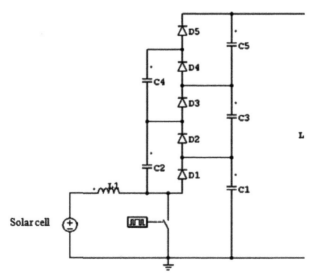

FIGURE 22.2   A dc–dc MLBC for three levels.

22.4   MAXIMUM POWER POINT TRACKING ALGORITHM

Figure 22.3 indicates that the characteristic output power curve for the solar cell shows the solar cell's work at its maximum power under a given temperature and irradiance, a MPPT control algorithm is employed to harvest this maximum power from the cell. As mentioned, it was said that many MPPT control algorithms have been proposed so far in the literature. The well-known algorithm called P&O has been employed in this work. Figure 22.3 depicts a flow chart explaining the main steps of it. It operates by perturbing the PV array voltage (i.e., incrementing or

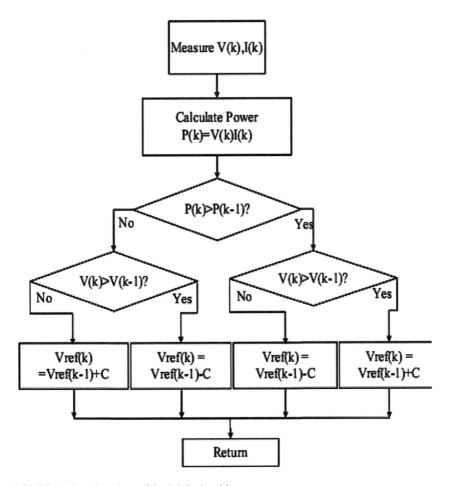

FIGURE 22.3   Flowchart of the P&O algorithm.

decreasing) and comparing the PV output power with that of the previous perturbation cycle. If the perturbation leads to an increase (decrease) in array power, the subsequent perturbation is made in the same (opposite) direction. In this manner, the peak power tracker continuously seeks the peak power condition. The output power of the solar cell in this algorithm is not constant but it oscillates around the maximum power point because there is no reference point in this algorithm. The algorithm is programmed using VHDL code, the inputs of the program are the PV current and voltage driven from two analog to digital converters (ADC), and the program output is the duty cycle that drives the gate of the boost converter switch. According to the flow chart, the program read the input signals (V, I), then calculate the PV module power by multiplying the two signals. The current values for the power, current, and voltage are compared with their counterparts of the previous values to execute the conditions. The duty cycle increment or decrement according to the change in the reference value is compared with the saw tooth carrier signal. The process repeats itself when the system oscillates at a certain point; this point is the maximum power point.

22.5   THE PROPOSED FPGA SYSTEM

22.5.1   THE CIRCUIT CONFIGURATION

Generally, the design of a digital controller impact is always the implementation of a suitable data acquisition path so that digital control requires particular care in signal conditioning and analog-to-digital conversion implementation.[4] A low-power prototype is built for experiment. Figures 22.4 and 22.5 show the overall hardware implementation circuit of the proposed system, that is, containing solar cell, multilevel boost converter boosts output voltage of the solar cell and at the same time extracts maximum power from it. There is one switch in the multilevel boost converter (MOSFET-16N60) and the fast recovery diode is FR605. All capacitors are 220 µF–400 V aluminum electrolytes and inductor has a value of 8 mH. MLBC extract maximum power by using Xilinx XC3S400 FPGA board, that is, containing the program of P&O; FPGA board read the PV output voltage and current by using a low-cost 8-bit analog-to-digital converter chip ADC0804lCN to convert the analog signal to digital signal.

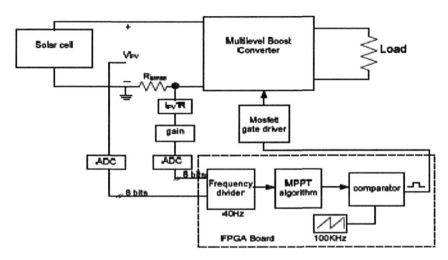

FIGURE 22.4   The overall hardware implementation.

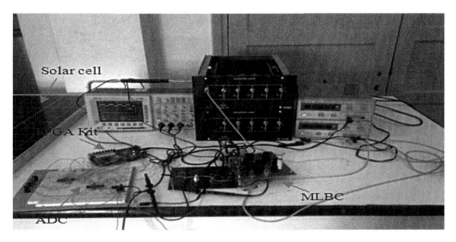

FIGURE 22.5   The experimental setup.

22.5.2   VOLTAGE AND CURRENT SENSING

The P&O method which is used to implement MPPT requires two sensors. Simple series and shunt resistances are used to measure the PV cell current and voltage, respectively. However, special care should be taken for the current sensor of this type, where most of the time it is easier and more reliable to measure voltage than current. Series sense resistor[11] is the conventional technique of sensing the current, it simply inserts a sense resistor in series with the return line of the PV, and the output current of the PV is determined by sensing the voltage across it. This method obviously acquires a power loss in Rsense. In this chapter, low resistance with high power rating resistance has been used (0.25 Ω, 5 W). The voltage across it equals 1.25 V when the maximum current (5 A) flows through it. But the full-scale range (FSR) of the ADC depends on the amplitude of the voltage reference of the ADC0804lCN (5 V) so if the maximum value input to the ADC equals 1.25 V, this means the FSR is not utilized. To boost this value to near the FSR, an amplifier circuit consists from operational amplifier is used as a gain which equals 3.

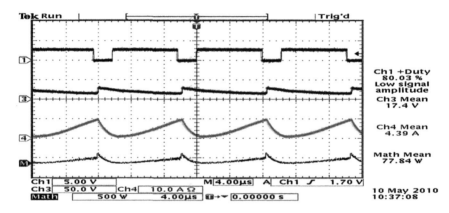

FIGURE 22.6   The duty cycle, output voltage, output current, and output maximum power, respectively, at time $T1$ during the day; $P = 77.84$ W.

22.5.3   EXPERIMENTAL RESULTS

The algorithm has been tested at different time during the day to ensure the testing is done for different environmental operation conditions. Figures 22.6–22.9 show the performance results obtained from the algorithm for different time during the day. These figures show different results for the variables due to the change of the environmental temperature. Figure 22.6 is done for maximum power at 77.84 W and then with small time difference, Figure 22.7 has been captured to show the capability of the control with tracking MPPT. For other different time with smaller insulation, Figures 22.8 and 22.9 are done where the maximum

extracted power is 75.89 and 66 W, respectively. Figure 22.10 shows the
PV voltage and current waveforms and how they reach the steady state
point for tracking the maximum power point. Figure 22.11 shows the
output voltage from MLBC and the pulses generated from FPGA board
that is used to switch the gate of MLBC at frequency 100 kHz. These
experimental setup proves the control algorithm and its function to track
the maximum power point.

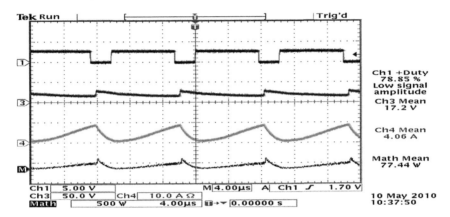

FIGURE 22.7   The duty cycle, output voltage, output current, and output maximum power,
respectively, at time $T2$ during the day; $P = 77.44$ W.

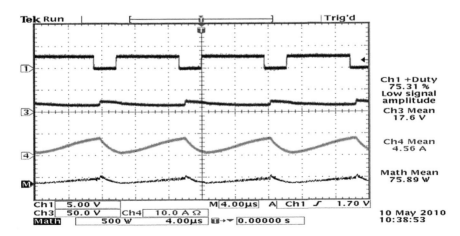

FIGURE 22.8   The duty cycle, output voltage, output current, and output maximum power,
respectively, at time $T3$ during the day; $P = 75.89$ W.

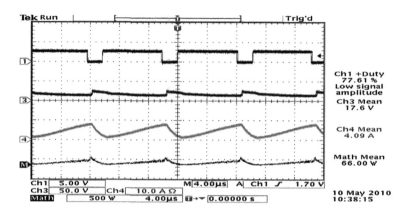

FIGURE 22.9   The duty cycle, output voltage, output current, and output maximum power, respectively, at time *T*4 during the day; *P* = 66 W.

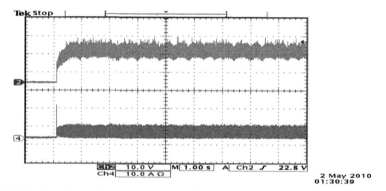

FIGURE 22.10   (See color insert.) The oscillation of the voltage and current waveform.

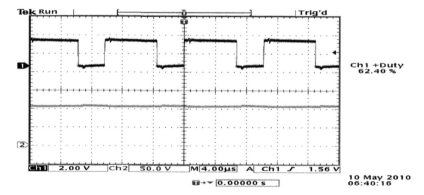

FIGURE 22.11   Recorded waveforms from Prototype II, 220 µF–400 V, L = 8 mH, V_{in} = 16 V, V_{out} = 110 V.

22.6 CONCLUSIONS

In this chapter, the implementation of the FPGA system to control the high-performance multilevel boost converter has been introduced. The implemented control has worked in the direction to track the maximum power from the PV source by using only one switch and implemented by XILINX FPGA that is considered as an efficient hardware for rapid prototyping. XILINX FPGA Web Pack software is used to generate PWM pattern by means of VHDL program. XILINX FPGA enables to make easy, fast, and flexible design and implementation. Also, a simple method of current sensing is used depending on the high performance of the FPGA with low resolution. Both simulation and experimental results are presented as confirmation of the approach presented.

KEYWORDS

- **photovoltaic**
- **FPGA**
- **multilevel boost converter**
- **maximum power point**
- **tracking**
- **simple sensor**
- **XILINX**

REFERENCES

1. Paiz, C.; Porrmann, M. In *The Utilization of Reconfigurable Hardware to Implement Digital Controllers: A Review*. IEEE International Symposium on Industrial Electronics, 2007; pp 2380–2385.
2. Al-Atrash, H.; Batarseh, I.; Rustom, K. In *Statistical Modeling of DSP-Based Hill-Climbing MPPT Algorithms in Noisy Environments*. IEEE Applied Power Electronics Conference and Exposition (APEC), 2005; pp 1773–1777.
3. Hua, C.; Lin, J.; Shen, C. Implementation of a DSP Controlled Photovoltaic System with Peak Power Tracking. *IEEE Trans. Ind. Electron.* **1998,** *45,* 99–107.
4. Yafaoui, A.; Wu, B.; Cheung, R. In *Implementation of Maximum Power Point Tracking Algorism for Residential Photovoltaic System*. 2nd Canadian Solar Buildings Conference, Calgary, June 10–14, 2007.

5. de Castro, A.; Zumel, P.; Garcia, O.; Riesgo, T.; Uceda, J. Concurrent and Simple Digital Controller of an AC/DC Converter with Power Factor Correction Based on an FPGA. *IEEE Trans. Power Electron.* **2003,** *18,* 334–343.

6. Khaehintung, N.; Wiangtong, T.; Sirisuk, P. In *FPGA Implementation of MPPT Using Variable Step-Size P&O Algorithm for PV Applications.* Symposium on International Communications and Information Technologies (ISCIT '06), 2006; pp 212–215.

7. Tsai, M.-F.; Hsu, W.-C.; Wu, T.-W.; Wang, J.-K. In *Design and Implementation of an FPGA-Based Digital Control IC of Maximum-Power-Point-Tracking Charger for Vertical-Axis Wind Turbine Generators.* Proceeding of the International Conference on Power Electronics and Drive System, PEDS, 2009.

8. Walker, G. Evaluating MPPT Converter Topologies Using a MATLAB PV Model. *J. Electr. Electron. Eng., Austr.* **2001,** *21* (1), 49–56.

CHAPTER 23

PROPOSAL FOR ECONOMIC IMPLEMENTATION OF PRECISION FARMING IN INDIA

D. DURGA BHAVANI*, MOUNIKA KAMATAM, and R. BHASHYA SRI BHARATI

Computer Science and Engineering, Nalla Malla Reddy Engineering College, Hyderabad, India

Corresponding author. E-mail: drddurgabhavani@gmail.com

ABSTRACT

Farming is a primary sector in India. In spite of the Green Revolution and usage of modern farming methods, farmers are not able to reap sufficient returns. The need of the hour is to maximize the productivity, and ensure that the farmers reap maximum benefit. In a country like India, where farming practices are largely dependent on monsoon, efficient and economic methods should be made available. This can be achieved using precision farming. The soil texture and properties vary from one location to another across the country. They also vary, within a farm of considerable size. There is a requirement to map and monitor various properties of the soil like mineral content, moisture levels, waterbed present underground, available nutrients, etc. A farmer can make a better choice, after knowing these details. It is a tough task to monitor different parameters manually. It will be easier if there is an automatic mechanism to do the same. We are concentrating on one single aspect that is moisture levels. The solution that we have come up with is a farming probe attached to a robot which moves around the field (crops arranged in matrix format) and uploads moisture values to the cloud. These values can be accessed via a smart phone. These moisture values can be mapped onto the field map so as to conditionally operate the irrigation system. In this chapter, an automatic machine for mapping moisture levels across the field has been proposed, using the principle of internet of things, for promoting the reduction of water usage in Indian agricultural practices.

23.1   INTRODUCTION

We can generally classify robot sensing into two modalities: remote contactless sensing (e.g., lasers, cameras, and ultrasound) and direct touch (e.g., haptics). Researchers have long speculated about a third sensing modality where "smart objects" or "smart environments" with embedded computation and sensing can directly measure and report salient information back to a robot. In more recent times, this general concept has garnered the moniker "Internet of Things."[1] The idea of robotic agriculture (agricultural environments serviced by smart machines) is not a new one. Many engineers have developed driverless tractors in the past but they have not been successful as they did not have the ability to embrace the complexity of the real world. Most of them assumed an industrial style of farming where everything was known beforehand and the machines could work entirely in predefined ways—much like a production line. The approach is now to develop smarter machines that are intelligent enough to work in an unmodified or seminatural environment. These machines do not have to be intelligent in the way we see people as intelligent but must exhibit sensible behavior in recognized contexts. In this way, they should have enough intelligence embedded within them to behave sensibly for long periods of time, unattended, in a seminatural environment, whilst carrying out a useful task. One way of understanding the complexity has been to identify what people do in certain situations and decompose the actions into machine control. This is called behavioral robotics and a draft method for applying this approach to agriculture is given in Blackmore.[2] Precision farming is defined as information and technology-based farm management system to identify, analyze, and manage variability within fields for optimum profitability, sustainability, and protection of the land resource.[3]

23.2   EXPERIMENTAL PART

Farming is a primary sector in India, which has developed over time; yet, it is not at par with the other countries. In spite of the Green Revolution and usage of modern farming methods, farmers are not able to reap sufficient returns. The need of the hour is to maximize the productivity and ensure that the farmers reap maximum benefit. In a country like India, where farming practices are largely dependent on monsoon, irrigation is a matter of concern for the farmers. It is necessary to shift to methods that ensure that sufficient amounts of water are provided at the right time in an efficient manner. The first step to attain that is to have a mechanism to monitor the moisture levels across the field (Table 23.1).

TABLE 23.1 Survey Results When Asked About Their Art of Agriculture in India.

Name	Designation	Remarks
Yadaiah	Farmer (4 ac)	I am currently using HYVs and I am happy with this yield
Bhaskar	Farmer (5 ac)	I cannot afford a robot
		I'd rather use cheap labor available
Dr. S. A. Hussain	Principal Scientist, PJTSAU	You have to convince the farmers that keeping track of these values can help in increasing their yields
		The robot should be affordable
Mr. V. Brijesh	John Deere	The idea is good and will be helpful in increasing the yield
		Try and map other values also
Mr. Krishna Reddy	Landlord (100 ac)	Keeping a track of these values for such a large area is not possible
Mr. Komaraiah	Farmer (10 ac)	I am not sure about how useful it will be to track the moisture levels
		My father got sick due to continuous exposure to chemicals
		If you can find a way to efficiently spray the chemicals, then we'll see
Mr. Damodar	Farmer (10 ac)	I will use it if it will be useful in cutting down my costs
Mr. Subramaniam	Professor: Agronomy	It will be useful if one such robot is bought collectively for 4–5 farms
		The farmers will need a lot of convincing to go for one such robot
Mrs. Sri Lakshmi	Professor: Agronomy	Not only monitoring the moisture levels, but also add a mechanism to spray the insecticide and should be robust

23.2.1 SURVEY TO UNDERSTAND THE PROBLEM

23.2.2 ARCHITECTURE

The device can comprise an Arduino microcontroller, servomotor, moisture sensor, motor driver, and ESP module (a low-cost Wi-Fi microchip) which can be connected to the cloud platform Thingspeak, an android application. This application can be used by the farmer to view the moisture levels at each and every plant, in a graphical format.[4]

Here, the moisture sensor is connected to the Arduino board, placed on the chassis of the robot. It is attached to a servomotor that facilitates in changing the angle. This moisture sensor measures the moisture level at each plant and sends the information to the Arduino board. A Wi-Fi module (ESP8266) provides internet connectivity to the robot. This module reads the data from the Arduino board and uploads the data into the cloud platform. We used the Thingspeak cloud platform. This platform provides mechanisms to store and view the data read by sensors. A companion application called Thingsview is used in the mobile phone by the farmer to view the data in the form of a graph (Fig. 23.1).[3]

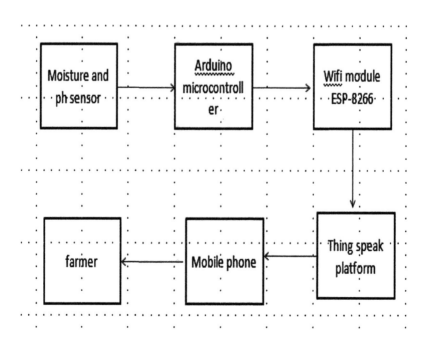

FIGURE 23.1   Technology architecture of the device.

25.2.3   KEY FEATURES

- **Compact**—The device can be designed to be small in size. This increases its maneuverability. It should not be bulky like other machines and should be easy to maintain.
- **Automatic collection and display of data**—The device should collect the data on its own by moving around the field.
- **No manual operation required**—The farmer has to put on the device and it should start performing its tasks. In the end, he can view the collected data. The farmer need not to stay with the device as it moves around the field.
- **Eco friendly, as it runs on electrical power**—The device should have rechargeable batteries that will make it ecofriendly when compared to the other machines used for farming, which run on fossil fuels.
- **Economic**—Farmers cannot afford costly products. The device should be affordable.[5]

23.3   CONCLUSION

Precision farming can be adopted into the Indian farming methods. There are certain geographical, economical, and technological constraints for the full-fledged usage of precision farming in India. A major drawback is the lack of awareness. The need of the hour is for economical and robust solutions. The prototype developed by us was able to achieve certain level of these constraints. A companion application can be further developed to keep track of multiple fields at once, to give weather reports, etc. Also, the prototype developed by us can be further improvised in such a way that it is capable of spraying fertilizers or insecticides autonomously. This reduces the risk that the farmers faced on being exposed to such harmful chemicals. In the long term, these types of solutions have the potential to better the lives of those who use them. Despite their focus on the farming production process, such applications and devices motivate users to embrace evolving communication skills and technologies by directly linking them to their livelihoods. Plus, a growing number of mobile tools are being developed to help Indian farmers scale up by going online to market and sell their products. Applications which help in connecting agricultural buyers with local farmers, thus eliminating middlemen who would reduce the farmers' margins—are quickly gaining traction. Going by the current pace, one can say that the opportunities are limitless.

KEYWORDS

- **robot sensing**
- **remote contactless sensing**
- **direct touch**
- **robotic agriculture**
- **driverless tractors**
- **smart environments**

REFERENCES

1. Wang, J.; Schluntz, E.; Otis, B.; Deyle, T. *A New Vision for Smart Objects and the Internet of Things: Mobile Robots and Long-Range UHF RFID Sensor Tags.*
2. Blackmore, S.; Stout, B.; Wang, M.; Runov, B. Robotic Agriculture Conference. European Conference on Precision Agriculture, 2005; pp 621–628.
3. Singh, A. K. *Precision Farming*; Water Centre Technology, IARI: New Delhi.
4. Van Briesen, J.; van de Venne, G. www.winchesterthurston.org/pge.cfm?p=3364.
5. Tali, D. www.good.is/articles/agricultural-apps-bridge-literacy-gaps-in-india.

CHAPTER 24

DEVELOPMENT OF SDC–SDF ARCHITECTURE FOR RADIX-2 FFT

G. DEESHMA VENKATAKANAKADURGA[*] and
G. R. L. V. N. SRINIVASARAJU

Department of ECE, SVECW, Bhimavaram, Andhra Pradesh, India

[*]*Corresponding author. E-mail: deeshma555@gmail.com*

ABSTRACT

In this chapter, we propose a single-path delay commutator (SDC)–feedback (SDF) architecture for Radix-2 fast Fourier transform (FFT) and presented its simulation results. The Radix-2 FFT architecture includes one SDF stage and $\log_2$ N-1 SDC stages. The SDC processing engine is used to attain 100% hardware blocks utilization by using the common arithmetic resources in the time multiplexed approach, including the architectures adders and multipliers. To get output sequence in normal order, we also have a bit reverser block; through this we can reduce 50% memory usage. The resultant architecture is simulated and design verification was done by using the software ModelSim and XilinxISE. The proposed architecture was achieved with reduced area and delay as well as improved performance.

24.1 INTRODUCTION

Fast Fourier transform (FFT) was proposed by Cooley and Tukey. From the last few years, digital signal processing has a wide range of applications. It is one of the important components in the field of digital signal processing, especially in the advanced communication systems such as orthogonal frequency division multiplexing, digital video broadcasting, and asymmetric digital subscriber line. FFT[4] plays vital role in these applications. To fulfill the requirements of these applications, FFT must occupy less silicon area, low power consumption, low latency, and high throughput.[1]

From the literature survey, the pipelined FFT architectures[2] are of three types. They are multipath delay commutator (MDC), single-path delay feedback (SDF), and single-path delay commutator (SDC). MDC architecture is used typically to process multiple-input data streams due to its high throughput rate. However, we can implement this through single-input data stream because it can utilize low hardware. SDF architecture[5] is the preferred solution to the single-input data stream, because the memory size required reaches the minimum and the multipliers are fully utilized. However, the utilization of adders is still very low. SDC architecture[6] is seldom used to process the single-input data stream, because it uses more memory resources than SDF and has a more complicated control.

Radix-2 FFT architecture mainly performs two operations. They are addition and subtraction. After completion of subtraction operation, it indeed involves complex multiplication.

In this chapter, we propose a combined SDC-feedback radix-2 FFT architecture[7]; it contains $\log_2 N - 1$ SDC stages, one SDF stage, and 1 bit reverser. The proposed architecture can generate the output sequences in same order as inputs.[3]

24.2   16-POINT RADIX-2 DIF–FFT ARCHITECTURE

We know that the radix-2 FFT is deduced from discrete Fourier transform (DFT) by dividing the N-point DFT into many two-point DFTs.[8] The data flow graph of 16-point radix-2 FFT is shown in Figure 24.1.

In Figure 24.1, we have 16 inputs and 16 outputs. We are applying the inputs in normal order but we didn't get the outputs in same order as inputs. From Figure 24.1, we notice that it has 16 paths.[9] So, to reduce these drawbacks, we implement SDC feedback architecture.

24.3   COMBINED SDC–SDF RADIX-2 FFT

24.3.1   PROPOSED FFT ARCHITECTURE

The proposed FFT architecture contains one prestage, $\log_2 N - 1$ SDC stages, one poststage, one SDF stage, and 1 bit reverser as shown in Figure 24.2a. The prestage separates the complex input data to a new sequence.[10]

That is real part followed by the corresponding imaginary part. The post stage changes back the new sequence to the complex format. The SDC stage contains an SDC PE; it can utilize 100% arithmetic resource including both complex adders and multipliers. The last stage is the SDF stage. It is similar to the radix-2 SDF, containing an adder and a subtracter.[11] By using the modified addressing method, the even index data are written into memory in normal order and these data are taken from memory in bit-reversed order while the odd-indexed data are written in bit-reversed order. Finally, we get the even data in normal order. So, bit reverser needs only $N/2$ data buffers.[12]

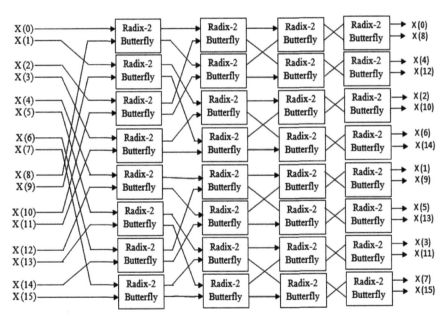

FIGURE 24.1 DFT of 16-point DIF–FFT.

24.3.2 SDC PROCESSING ENGINE

In Figure 24.2b, the SDC processing engine contains a data commutator, a real add/subunit, and an optimum complex multiplier unit. To reduce the arithmetic resource of the SDC PE, the important factor is to increase the utilization of the arithmetic components via changing the above three units' data sequences.

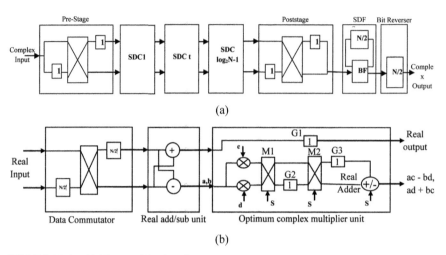

FIGURE 24.2   (a) The combined SDC–SDF architecture for Radix-2 FFT. (b) The SDC PE for Radix-2 FFT.

In each SDC stage, the data commutator changes input data into a new data sequence and their index difference is $N/2^t$, where t indicates the index of the stage. For each of the input data, addition and suboperations are performed through real add/subunit.

The output data sequence of optimum complex multiplier unit and the real add/subunit should be same. Finally, its output sequence is also the output sequence of the SDC stage t and its input sequence to the SDC stage is $t + 1$.

24.3.3   OPTIMUM COMPLEX MULTIPLIER UNIT

In Figure 24.2b, optimum complex multiplier unit consists of two multiplexers, 1.5 word memory, two real multipliers, and one real adder. The signal s controls the addition and subtraction operations of real adder.[13]

For the input data sequences (0_r, 8_r) and (0_i, 8_i) at the real add/subunit, the output of addition part 0_r and 0_i will directly pass to the delay memory to generate a new sequence 0_r* and 0_i* with one cycle delay in consecutive two cycles, while the output of difference parts 8_r and 8_i goes to the real multipliers to generate (c × 8_r, d × 8_r) and (c × 8_i, d × 8_i) before reordering.

The above process can be applied to the other couples in the stage 1, for example, (2_r, 10_r) and (2_i, 10_i), and so on. If we perform the above process through $\log_2 N - 1$ SDC stages to completion, then the majority part of the radix-2 FFT[14] computation will be completed.

24.4   RESULTS AND COMPARISON

The design of combined SDC–SDF architecture for radix-2 FFT has been made by using Verilog Hardware Description Language. The simulation results have been evaluated by using ModelSim 6.3c and synthesis performances are estimated by using Xilinx ISE 14.1.[15]

In Figure 24.3a and b, complex input is the combination of real part and imaginary part. Here, in_real is the real part and in_imag is the imaginary part. We are applying 16 inputs (complex) of 32-bit range through single path, clk, control as well as twiddle of 3 bit. Signal s of 4 bit represents number of inputs.[15]

In Figure 24.3c and d, complex output consists of real part and imaginary part. Here, out_real is the real part and out_imag is the imaginary part. After receiving 16 inputs (complex data), the results (out_real and out_imag) are obtained through single path of 32-bit range.

In Figure 24.4, register-transfer level (RTL) schematic shows input and output signals. clk, in_real of 32 bit, in_imag of 32 bit, control of 5 bit, twiddle of 3 bit, and s are inputs. out_real of 32 bit and out_imag of 32 bit are outputs.[16]

In Figure 24.5, detailed view shows one prestage, three SDC stages, one poststage, one SDF stage, and 1 bit reverser. Whatever the components used in the modules are visible in the detailed view of the RTL schematic. The design summary of SDC-feedback radix-2 FFT is shown in Table 24.1. Number of slice registers 3%, number of slice look up tables (LUTs) is 72%, number of digital signal processing (DSP) 48E 1s is 25%, minimum clock period is 22.863 ns, frequency is 43.739 MHz, and maximum combinational path delay is 2.375 ns.

TABLE 24.1   Design Summary of Single Path Delay Commutator-Feedback radix-2 FFT.

S. No.	Parameters	Value
1.	No. of slice registers (%)	3
2.	Number of slice LUTs (%)	72
3.	Number of DSP 48E 1s (%)	25
4.	Number of bonded IOBs (%)	63
5.	Minimum clock period (ns)	22.863
6.	Frequency (MHz)	43.739
7.	Maximum combinational path delay (ns)	2.375

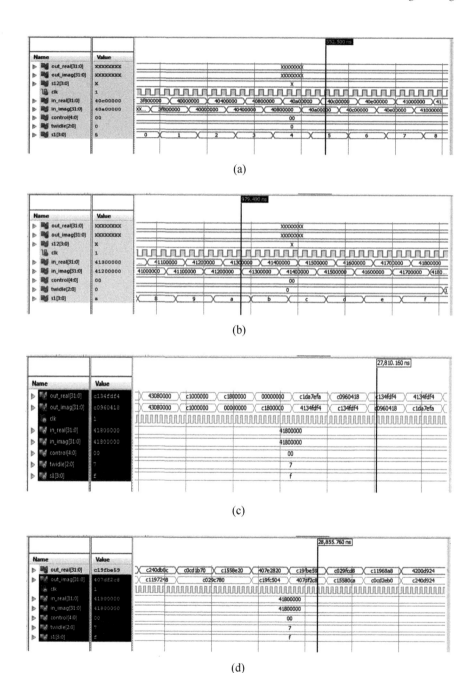

FIGURE 24.3 (a) Simulation Waveform 1 of radix-2 SDC–SDF FFT; (b) simulation Waveform 2 of radix-2 SDC–SDF FFT; (c) simulation Waveform 3 of radix-2 SDC–SDF FFT; and (d) simulation Waveform 4 of radix-2 SDC–SDF FFT.

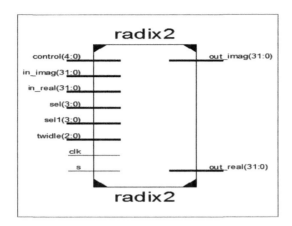

FIGURE 24.4   RTL Schematic of radix-2 SDC–SDF FFT.

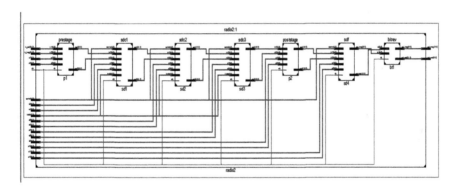

FIGURE 24.5   Detailed view of radix-2 SDC–SDF FFT.

SDC-feedback radix-2 FFT architecture[17] is compared with 16-point radix-2 decimation in frequency (DIF)–FFT for various parameters like number of bonded input/output blocks (IOBs), number of slice LUTs, number of DSP 48E 1s, maximum combinational path delay, maximum frequency, and minimum clock period. The implementation results give the same outputs, but delay and area are less compared with 16-point radix-2 DIF–FFT.[18] Table 24.2 shows that all the parameters of SDC–SDF radix-2 FFT are having better result than 16-point radix-2 DIF–FFT.[19]

TABLE 24.2   Comparison Between Single-path Delay Commutator-Feedback Radix-2 FFT and 16-Point Radix-2 DIF–FFT.

S. No.	Parameters	SDC–SDF radix-2 FFT	16-point radix-2 DIF–FFT
1.	Number of slice LUTs (%)	72	75
2.	Number of DSP 48E 1s (%)	25	34
3.	Number of bonded IOBs (%)	63	720
4.	Max. combinational path delay (ns)	2.375	102.169
5.	Min. clock period (ns)	22.863	74.239
6.	Frequency (MHz)	43.739	13.470

24.5   CONCLUSION

The proposed SDC-feedback (SDC–SDF) radix-2 FFT architecture produces the output data in the normal order. The proposed architecture reduces number of complex multiplications, additions, and number of stages compared with the general butterfly architecture. The SDC-feedback radix-2 FFT architecture is simulated using ModelSim and design verification; area and delay reports are generated using Xilinx ISE 14.1.

It is observed that in the proposed architecture, the device utilization is also much reduced, which in turn reduced the area. The number of slice LUTs utilized in the existing system is 75% while that of SDC–SDF radix-2 FFT architecture is 72%. This indicates that the proposed architecture could effectively reduce the area consumption. Similarly, the delay of the existing method is 102.169 ns, while the delay of our architecture is 2.375 ns. This proves the efficiency of our architecture both in terms of delay and area.

KEYWORDS

- Radix-2 FFT
- single-path delay commutator processing engine
- single-path delay feedback
- multipath delay commutator
- ModelSim

REFERENCES

1. Cheng, C.; Parthi, K. K. High Throughput VLSI Architecture for FFT Computation. *IEEE Trans. Circuits Syst. II, Exp. Briefs* **2007,** *54* (10), 339–344.
2. Wold, E. H.; Despain, A. M. Pipeline and Parallel-Pipeline FFT Processors for VLSI Implementation. *IEEE Trans. Comput.* **1984,** *C-33* (5), 414–426.
3. Chang, Y. N. AN Efficient VLSI Architecture for Normal I/O Order Pipeline FFT Design. *IEEE Trans. Circuits Syst. II, Exp. Briefs* **2008,** *55* (12), 1234–1238.
4. Liu, X.; Yu, F.; Wang, Z. K. A Pipelined Architecture for Normal I/O Order FFT. *J. Zhejiang Univ. Sci. C* **2011,** *12* (1), 76–82.
5. Cortes, A.; Velez, I.; Sevillano, J. F. Radix r^k FFts: Matricial Representation and SDC/SDF Pipeline Implementation. *IEEE Trans. Signal Process.* **2009,** *57,* 2824–2839.
6. Wang, Z.; Liu, X.; He, B.; Yu, F. A Combined SDC–SDF Architecture for Normal I/O Pipelined Radix-2 FFT. *IEEE Trans. Very Large Scale Inegr. (VLSI) Syst.* **2015,** *23* (5), 973–977.
7. Cimini, L. J. Analysis and Simulation of a Digital Mobile Channel Using Orthogonal Frequency Division Multiplexing. *IEEE Trans. Commun.* **1985,** *33* (7), 665–675.
8. Tang, S. N.; Tsai, J. W.; Chang, T. Y. A 2.4-GS/s FFT Processor for OFDM-Based WPAN Applications. *IEEE Trans. Circuits Syst. II, Exp. Briefs* **2010,** *57* (6), 451–455.
9. Jung, Y.; Yoon, H.; Kim, J. New Efficient FFT Algorithm and Pipeline Implementation Results for OFDM/DMT Applications. *IEEE Trans. Consum. Electron.* **2003,** *49* (1), 14–20.
10. Shin, M.; Lee, H. A High-Speed, Four-Parallel Radix-2^4 FFT Processor for UWB Applications. In *Proc. IEEE ISCAS,* May 2008; pp 960–963.
11. Sansaloni, T.; Perez-Pascual, A.; Torres, V.; Valla, J. Efficient Pipeline FFT Processors for WLAN MIMO-OFDM Systems. *Electron. Lett.* **2005,** *41* (19), 1043–1044.
12. Oh, J. Y.; Lim, M. S. Area and Power Efficient Pipeline FFT Algorithm. In *Proc. IEEE Workshop Signal Process. Syst. Design and Implementation,* Nov. 2005; pp 520–525.
13. Cho, T.; Tsai, S.; Lee, H. A High-Speed Low-Complexity Modified Radix-2^5 FFT Processor for High Rate WPAN Applications. *IEEE Trans. Very Large Scale Inegr. (VLSI) Syst.* **2013,** *21* (1), 187–191.
14. Garrido, M.; Grajal, J.; Sanchez, M.; Gustafsson, O. Pipelined Radix-2^k Feedforward FFT Architectures. *IEEE Trans. Very Large Scale Inegr. (VLSI) Syst.* **2013,** *21* (1), 23–32.
15. Yang, L.; Zhang, K.; Liu, H.; Huang, J.; Huang, S. An Efficient Locally Pipelined FFT Processor. *IEEE Trans. Circuits Syst. II, Exp. Briefs* **2006,** *53* (7), 585–589.
16. Lenart, T.; Owall, V. Architectures for Dynamic Data Scaling in 2/4/8k Pipeline FFT Cores. *IEEE Trans. Very Large Scale Inegr. (VLSI) Syst.* **2006,** *14* (11), 1286–1290.
17. Ayinala, M.; Brown, M.; Parthi, K. Pipelined Parallel FFT Architectures via Folding Transformation. *IEEE Trans. Very Large Scale Inegr. (VLSI) Syst.* **2012,** *20* (6), 1068–1081.
18. Bi, G.; Jones, E. V. A Pipelined FFT Processor for Word-Sequential Data. *IEEE Trans. Acoust. Speech Signal Process* **1989,** *37* (12), 1982–1985.
19. Gold, B.; Rader, C. M. *Digital Processing of Signal*; McGraw-Hill: New York, NY, 1969 (Chapter 6).

CHAPTER 25

ADVANCED TOUCH SCREEN SYSTEM FOR ELDERLY PEOPLE

M. N. S. LAHARI* and K. UMAPATHY*

Department of ECE, Sri Chandrasekharendra Saraswathi Vishwa Mahavidyalaya, Kancheepuram, Tamil Nadu, India

Corresponding author. E-mail: mns.lahari@gmail.com; umapathykannan@gmail.com

ABSTRACT

People age with time. Everything that we do is limited by a boundary called time. But with the help of technology, household work could become simpler for the elderly. The present work is an introduction to a touch control-based system, which is all about remaining stationary at a place and controlling ponderous electric weights. This is possible by just touching a surface of a device placed in the hands of the elderly. Often, there are disputes in families regarding elderly people's work to be done, especially in a home where everybody is busy earning. Though there is already a system that is framed with several control buttons, its practical applications are not imminent. This could be because when we have several options, there is always scope for confusion. This work does not just require a touch screen device, but, with further modifications, it could be implemented in our smartphones, which we hold every minute of the day. The working of this touch control-based system not just fleeces the complexity of the existing system, but also enables limiting access by an unauthorized person, which is a highly secure feature. In order to shorten the description of the system, it could also be put in this way like affordable, user-friendly, eminent, and portable.

25.1   INTRODUCTION

Life has become easy with advancements in science and technology. Every-thing seems to be so redundant in front of advancing technology. However, all these have proved to be nothing to people who stay in their small world. Elderly people, who often become aged and helpless, are certain times left at home with nothing to do. Studies show that 80% of the elderly people become bedridden or paralyzed because they stay at homes with no proper motivation to move forward in their lives. Besides this, basic health issues, like arthritis, lack of vision, etc., seem to become a huge hindrance to their mobility.

Someone has to give them ease, and maybe that is the only reason why youngsters have initiated a step to introduce technology to the elderly people and make their life more cheerful. When we discuss touch control-based systems, it is all about controlling the electric devices placed inside our homes, all of which usually require a controlling device. Considering it as an advantage, a touch-based control system has been proposed in this chapter. When we describe it as portable and user-friendly, we mean that it eliminates the issues arising from movement of the user from his or her place where he or she could just sit and enjoy the pleasure of electronic world all around them. It also enables the long-range accessibility of the devices in the absence of other family members.[2]

All that this system requires is a touch screen module, a microcon-troller, and an RF (radio frequency) wireless technique. This combination of a microcontroller and an RF has many times proven to be the most advantageous way of improving the existing technology. In addition, this combination also provides an easy way for designing the system. Since it is entirely a touch-based system, it could be accessed by any person (including the children in the house). To overcome this confusion, there is a limit for accessing the system, in that only an authorized person can get access to the system.[3] Moreover, the most significant fact is that the implementation of touch-detecting sensors and also the microcontroller, which are easily available in the market, involves low cost and low-power consumption.[4]

25.2   PREVAILING METHODS AND THEIR OUTCOMES

It cannot be claimed that the method which is being introduced is the first and most important method ever. No, it is not at all so. Several other methods that already exist in the world of technology lay their path over fields of communication like network and management, power line, global system for mobile communication (GSM), Bluetooth, etc., which could only explain the better working but not the security and affordability of the system.

25.2.1   NETWORK AND MANAGEMENT

When the software meddles up with the electronic working, it increases the complexity as well as the cost of the devices for the user. Similarly, when Java is incorporated to build up the network security features, it not just increases the wire installation factors but also the user guidelines seem to become highly complicated for common people. The most important disadvantage of implementing this system is that it is susceptible to duplicity, lacks flexibility and there are more chances for causing disturbances at both input and output signals.

25.2.2   POWER LINES

Use of the power line communication limits the access of the device within the user residential areas, which could actually become complicated for users who are physically unfit.

   In cases where the dorm is placed near an electric field case, the noise disturbances generated by it could be completely unstable for the user.[5]

25.2.3   GSM

As the name suggests, it is about the radiating frequencies through several different bandwidths which make it possible for creating confusions in the developed system. Not just that, as it deals with tuning out a perfect frequency for signal transmission, the cost linked with the system cannot be afforded by everyone. Hence, it cannot be employed in a system like this.

25.2.4   BLUETOOTH

As the name suggests, it is totally dependent on sharing things or involvement of two or more people with regard to single information. This fleeces the security system of the device, increasing the complexity as well as the economic condition of the system that has to be employed, and the drawback in the system is sharing the information to several other systems that could easily be controlled within and around.[4]

25.3   WORKING OF THE PROPOSED SYSTEM

As per the above mentioned systems and their drawbacks, it could all be removed or overcome through implementation of RF communication methodology. It comprises a transmitter and a receiver, which are employed to ensure the efficiency of receiving and transmitting the required signal. The execution of this device through RF communication has been carefully carried out through touch screen modules for the purpose of security as well as to control the electronic devices by user.[2]

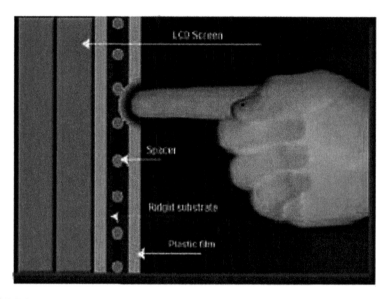

FIGURE 25.1   Inner layers of the proposed touch screen.

25.3.1   *RESISTIVE TOUCH SCREENS FOR THE DEVICES*

The word resistive indicates the type of material bound in devices. They are special sort of devices generally found in electrical engineering streams, which are also known as sensitive touch screen displays. These are internally composed of multilevel flexible sheets that are generally coated using the resistive material and further separated by a thin air gap or may be the spacing of even a microorganism. The first layer is said to consist of electrodes on substrates like glass and plastics that form a screen of thick fibers, which are highly resistive and sensitive to human touch. It is said to have a very high resolution that has the ability for determining the accurate touch. The most important fact of this sort of technology is that it could determine the touch even through well-equipped gloves on the hands.[1]

Its working principle in the output of our system is analogous, which has to be converted into digital for further transmission to a processor/controller for a better outcome as per the layout shown below.

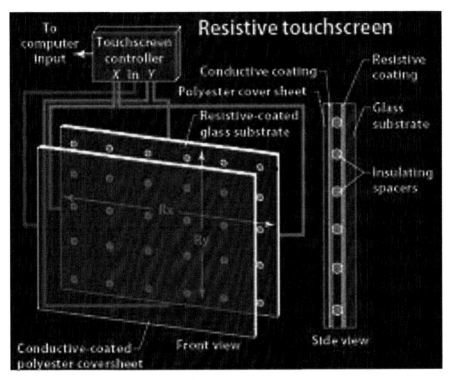

FIGURE 25.2   Touch screen layout diagram.

As far as the IC in the proposed system is concerned, the IC TSC2020, which has special features like low-cost, resistive touch screen design, and several other handheld applications, has been used. The functional block diagram shows a complete ultra-low-power 12-bit analog-to-digital converter that is said to consist of both sensor drivers and a control logic to measure the touch pressure. Unlike any other touch-based controllers, this TSC2020 is capable of accepting a maximum of three touches simultaneously and delivering a very low power standby touch detection. The only condition that is verified here is identifying the simultaneous touch of the user and generating low-power consumption signals.

25.3.2   PIC MICROCONTROLLER

PIC16F628A is utilized in the proposed system, which controls the entire system just by being in the form of a small man-made chip. The coding for the controller is burnt by writing the source code that enables control over the working of the proposed system. The coding used here is version 4 of Keil software. It also enables the development of controller by graphically further helping it to generate a block diagram explanation for the working of the system. When the controller receives input from the screen, it activates the corresponding action that has to be performed. PIC is known to enhance efficiency in generating results at a faster rate when compared to the controllers.

And also that the PIC controller is considered to be the most compatible controller for encoding and decoding of received analog to the digital information from the user. This method of usage results in easy creation of codes or the programs by simply selecting and dropping of the required programs onto the block diagram. Hence, the developed program is now being burnt into the PIC16F628A using the PIC burner.[1]

25.3.3   RF WIRELESS COMMUNICATION

RF refers to the frequencies that fall into the electromagnetic spectrum associated with radio wave propagation. And when the process is said to reoccur with the antenna, it generates electromagnetic radiations that propagate in the applied field through space. As it is already known that the wavelength is inversely proportional to frequency, the wavelength of the electromagnetic wave thus radiated is inversely proportional to RF frequency. The present module works as a remote based on the principle of frequency modulation of the given signal at 433 MHz, which could be accessed at a range of 250 m outdoor and 450 m indoor which are internally connected to the electronic gadgets or the devices through an electric relay, further accommodated with an RF decoder. An input of 433 MHz is easily triggered using a 4-bit chip, which has a very small and simplified circuit involving no greater risks. From several other aspects, considering 433 MHz is highly advantageous for a system that tends to operate in an efficient way at short ranges.

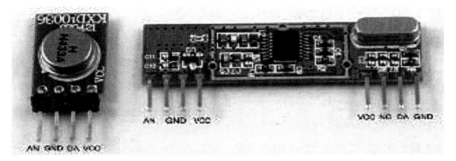

FIGURE 25.3 Transmitter and receiver.

When we carefully look at the circuit, it can be seen that the antenna provided at the top receives signals from transmitter. After being encoded, it is sent to the receiver end and is further decoded. There is an extension for the microcontroller with an electronic relay as it is known to have an ability to cause mobility in any sort of electric devices (only when the control key is held pressed). The data signal extracts the information from the RF receiver end through carrier frequency. An LED is placed at the circuit in the receiver's end for regulating the current status. As per the decoded information at the decoder, the output signal is set *high* or *low*. When signal is *high*, it drives the relay through ULN2003 which is a transistor array.

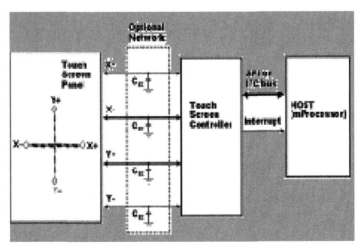

FIGURE 25.4 Block diagram of proposed network.

25.4   DESIRED ALGORITHM

Step I: Start the program and finish the step of authentication by registering the user.

Step II: If authentication is a failure, return to step I. Else unlock the system for controlling electric loads.

Step III: Initialize I/O ports.

Step IV: Wait for the input from user by setting a time using a timer. If no input is received within the time, then activate standby mode.

Step V: Else read the coordinates and activate the touch screen module.

Step VI: Change the status of system working based on the coordinates.

That is, if coordinates are of II quadrant switch on the load 1

  else coordinates are of I quadrant switch on load 2

  else coordinates are of III quadrant switch on load 3

  else coordinates are of IV quadrant switch on load 4

End the loop.

Step VII: Perform the work that is desired and move to step V for the next coordinate.

Step VIII: Lock the command as per the user's press.

Step IX: Else lock the module as per the time being set at the timer automatically.

Step X: Go to step I for restarting the system.

25.5   CONCLUSION AND SCOPE OF DEVELOPMENT

The sort of technology discussed here is under the practicality which is dealing with the perfect outputs that are expected at a single touch. To generate them, a 2×2 matrix is proposed by separating the touch screen into four quadrants as per the coordinate system. The present system is developed and tested for a distance of 100 m using a 4-bit digit codes. And also, the use of basic devices like microcontrollers, timers, relay, and a touch screen modules makes it more amiable and eminent. As per the above work and algorithm designed, using several other modifications, this system could be implemented in mobile phones. It is a low-cost and more profitable system designed to ensure that there are no more loopholes in the proposed system.

KEYWORDS

- **elderly people**
- **touch-based control**
- **electric loads**
- **affordable**
- **portable**
- **eminent**
- **user-friendly**

REFERENCES

1. Mainaridi, E. In *Density Based Resistive Touch Screen System*. 4th IEEE Conference, Aug 2008; p 34.
2. Gill, K. In *Zigbee Based Control System*. IEEE Transaction on Consumer Electronics, Sept 2012.
3. Lee, S.-W.; Kim, Y.-J.; Lee, G.-S.; Cho, B.-O.; Lee, N.-H. In *A Remote Behavioral Monitoring System for Elders Living Alone*. Proceedings of International Conference on Control, Automation and Systems, Oct 2007; pp. 2725–2730.
4. O'Farrell, M. J. Low Level Radio Frequency Control of RIA Superconducting Cavities. Master of Science Thesis, Department of Electrical and Computer Engineering, Michigan State University, 2005.
5. Khiyal, M. S. H.; Khan, A.; Shehzadi, E. SMS Based Wireless Home Appliance Control System (HACS) for Automating Appliances and Security. *Issues Inf. Sci. Inf. Technol.* **2009,** *6,* 887–894 (Software Engineering Dept., Fatima Jinnah Women University, Rawalpindi, Pakistan).

CHAPTER 26

APPLICATIONS OF MICROCONTROLLERS

NIHAR RANJAN PANDA[1*], P. N. S. SAILAJA[1], and
RUPALI SATAPATHY[2]

[1]*Department of ECE, SITAM, Vizianagaram, Andhra Pradesh, India*

[2]*Department of EEE, HIT, Bhubaneswar, Odisha, India*

Corresponding author. E-mail: nihar.mits@gmail.com

ABSTRACT

Microcontrollers are applicable to a wide range of information processing tasks, ranging from general computing to real-time monitoring systems. The microcontrollers facilitate new ways of communication and how to make use of the vast information available online and offline both at home and in workplace. Most electronic devices—including everything from computers, remote controls, washing machines, microwaves, and cell phones to iPods and more—contain a built-in microcontroller. Microcontrollers are at the core of personal computers, laptops, mobile phones, and complex military and space systems. This work presents the general application of microcontrollers.

26.1 INTRODUCTION

A microcontroller is usually a silicon chip that contains millions of transistors and other components that process millions of instructions per second integrated with memory chips and other special purpose chips and directed by software.[3,4] It is a multipurpose, programmable microchip that uses digital data as input and provides results as an output once it processes the input according to instructions stored in its memory. Microcontroller use sequential digital logic as they have internal memory and operate on numbers and symbols represented in the binary numeral system. They are designed to perform arithmetic and logic operations that make use of data on the chip. General-purpose microcontroller in PCs is used for multimedia display, computation, text editing, and communication. Several Microcontrollers are part of embedded systems. These embedded microcontrollers provide digital control to automatically controlled products and devices, such as automobile engine control systems, implantable medical devices, remote controls, office machines, appliances, power tools, toys, and other embedded systems. Microcontrollers are designed for embedded applications, in contrast to the microprocessors used in personal computers or other general purpose applications consisting of various discrete chips. A typical example is shown in Figure 26.1.[1]

FIGURE 26.1 The die from an Intel 8742, an 8-bit microcontroller.

The development of the first microcontroller began in 1971; the Smithsonian Institution credits TI engineers Gary Boone and Michael Cochran with the successful creation of the first microcontroller in 1971. Surprisingly, this exceptional breakthrough in the field of electronics and communication was rather given a mundane name of TMS1802NC; however, the device wasn't ordinary. It had 5000 transistors providing 3000 bits of program memory and 128 bits of access memory. So, it was possible to program it to perform a range of functions. Meanwhile, Atmel and Peripheral Interface Controller (PIC) family developed the different microcontroller.

The function of the microcontroller is best described in a small computer on a single integrated circuit (IC) containing a processor core, memory, and programmable input/output peripherals. A modern microcontroller can complete this three-step process millions of times in 1 s.[1,5]

Microcontroller may be classified by their hardware architecture. The two basic types of hardware are complex instruction set computer (CISC) and reduced instruction set computer (RISC). CISC processors can perform complex functions with one instruction while RISC chips usually need multiple instructions. The (8051 controller) MCS family is based on the CISC architecture, while Atmel and PCI family chips are RISC systems.[12]

The following are the examples of microprocessor: MSP430F5x/6x, MSP430G2x/i2x, Microchip, Tenco Technology Co, United Sources Industrial, Shenzhen Guangfasheng Technology, Shenzhen Jiubaba Electronics, Atmel, Intel, *Advanced RISC Machines* (ARM), Texas Instrumentation, Rabbit, NXP Semiconductors, STMicroelectronics, Toshiba. Each of these microcontrollers has their versions and kinds.

The integration drastically reduces the number of chips and the amount of wiring and circuit board space that would be needed to produce equivalent systems using separate chips. Furthermore, on low pin count devices in particular, each pin may interface to several internal peripherals, with the pin function selected by software. This allows a part to be used in a wider variety of applications than if pins had dedicated functions.

Microcontrollers shine in situations where limited computing functions are required within an easily definable set of parameters. Microcontrollers excel at the low-grade computational functions required to run devices such as electronic parking meters, vending machines, simple sensors, and even home security equipment. Microcontrollers surround most Americans in their homes and offices, being present in devices such as televisions, remote controlled stereos, and even the digital computer components of a timer on a newer stove.

Microcontrollers have proved to be highly popular in embedded systems. Microcontrollers have numerous applications. Some examples of their simple applications are in (1) peripheral controllers of a computer such as the keyboard controller, printer controller, laser printer controller, LAN controller, and disk drive controller; (2) communication systems like numeric pagers, cellular phones, cable TV terminals, FAX and transceivers with or without an accelerator, video game, etc.; (3) biomedical instruments like an ECG LCD display cum recorder, blood cell recorder cum analyzer, and patient monitor system; (4) instruments such as an industrial process controller and electronic smart weight display system; (5) a target tracker; (6) an automatic signal tracker; (7) accurate control of the speed and position of a DC motor; (8) a robotics system; (9) a computer numerical control (CNC) machine controller; and (10) automotive applications like a close loop engine control, a dynamic ride control, an antilock braking system monitor, etc.[4,9]

The microcontrollers are classified in terms of internal bus width, embedded microcontroller, instruction set, memory architecture, IC chip, or very-large-scale integration (VLSI) core (VHDL or Verilog) file and family. There are 8-, 16-, and 32-bit microcontrollers.[6] For the same family, there may be various versions with various sources. The processors in microcontrollers are either general processors or purpose built.

26.2 MICROPROCESSOR

Microprocessor is a clock-driven semiconductor device consisting of electronics logic circuits manufactured by using large-scale integration (LSI) or VLSI technique. It is a computer processor which incorporates the functions of a computer's central processing unit on a single IC. Microprocessors use sequential digital logic as they have internal memory and operate on numbers and symbols represented in the binary numeral system. The microprocessors are performing arithmetic and logical operation that make use of data on chip. Thousands of items that were traditionally not computer related include microprocessors. These include large and small household appliances, cars (and their accessory equipment units), car keys, tools and test instruments, toys, light switches/dimmers and electrical circuit breakers, smoke alarms, battery packs, and hi-fi audio/visual components (from DVD players to phonograph turntables). Such products as cellular telephones, DVD video system, and HDTV broadcast systems fundamentally require consumer devices with powerful, low-cost microprocessors. Increasingly stringent pollution control standards effectively require automobile manufacturers to use microprocessor engine management systems, to allow optimal control of emissions over widely varying operating conditions of an automobile. Nonprogrammable controls would require complex, bulky, or costly implementation to achieve the results possible with a microprocessor.

The task of the microprocessor is described in preeminent way. They are fetching, processing, and decoding. In the fetching step, it gets an instruction from the computer's memory. In the decoding step, it decides what the instruction means. The last step is the processing itself, which involves the microprocessor's carrying out or performing the decoded set of instructions. A modern microprocessor can complete this three-step process millions of times in 1 s.

The following are examples of microprocessor: Intel, AMD, Elbrus, Fairchild Semiconductor, Motorola, Hewlett-Packard, IBM, Zilog Z80, ARM DEC, MIPS Technologies, National Semiconductors, NEC, NXP (Phillips), SPARC, Texas, and VIA.[1,4]

26.3   APPLICATION OF MICROCONTROLLERS

There are so many applications like peripheral controllers of a computer such as the keyboard controller, printer controller, laser printer controller, LAN controller and disk drive controller, house hold devices; communication systems like numeric pagers, cellular phones, cable TV terminals, FAX and transceivers with or without an accelerator, video game, etc.; biomedical instruments like an ECG LCD display cum recorder, blood cell recorder cum analyzer, and patient monitor system; instruments such as an industrial process controller and electronic smart weight display system; a target tracker, an automatic signal tracker; accurate control of the speed and position of a DC motor; a robotics system; a CNC machine controller; and automotive applications like a close loop engine control, a dynamic ride control, an antilock braking system monitor.[1,5]

26.3.1   HOUSEHOLD DEVICES

A home appliance control system is a system which provides various services to remotely operate on home appliances. They demand sophisticated, feature-rich products that are reliable and easy to use. Advanced motor control features for safe, quiet operation and use of green, power-efficient technology, as well as energy measurement, and control through connectivity with smart metering networks are the part of home appliance control system. Advance human–machine interface supports through touch screen technology[8] for a rich, easy user experience. Some home items that contain microcontroller include refrigerator, washing machine, motor controller, safety critical touch interfaces, televisions, microwaves, stoves, clothes washers, stereo systems, hand-held game devices, video game systems, dishwashers, home lighting systems, and even some refrigerators with touch screen digital temperature control.

The following picture (Fig. 26.2) gives an overview of how this system is going to work.

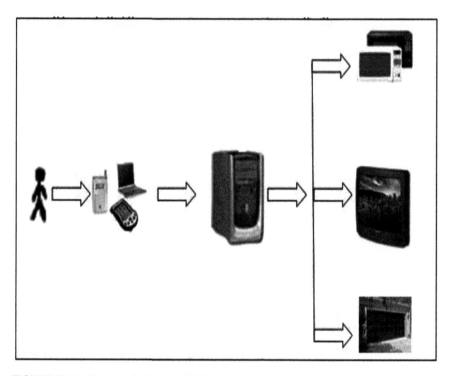

FIGURE 26.2   Diagram for household devices.

26.3.2   COMMUNICATION SYSTEM

Microcontrollers designed for communication applications include sections for handling communication protocols such as Wi-Fi, Bluetooth, ZigBee, CAN bus, infrared, USB, and Ethernet. Communication microcontrollers can be found in wireless devices and in wired network devices such as those in automotive applications. In Figure 26.3, it has shown how microcontroller can be connected to the mobile by using Bluetooth. In this example, we can control a relay by using mobile phone with the help of microcontroller.

The use of microprocessor in television, satellite communication, has made teleconferencing possible. Railway reservation and air reservation system also use this technology. LAN and WAN can be used for communication of vertical information through computer network.[11]

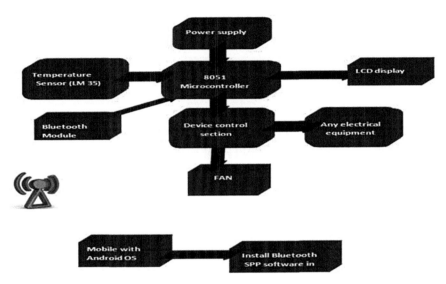

FIGURE 26.3   Mobile phone connection to microcontroller using Bluetooth.

26.3.3   AUTOMATIC PROCESS CONTROL

The term process control implies the technique of having precise control on any sequential process. Basically, any automatic process technique inputs from a process/machine (e.g., input from a sensor) or from switching command (e.g., input from a push switch or toggle switch) and uses outputs to control that process/machine according to the program (instruction) stored in the controller. The controller is capable of storing instructions, such as sequencing, timing, counting, arithmetic operation, data manipulation, and communication to control industrial machines and processes. Figure 26.4 illustrates a conceptual diagram of a process control.

For example, a system of a number of boilers supplied by a main water tank has been proposed. The water level in the main tank is controlled by a water level sensor; each boiler has two pipes, one is inlet and other one is outlet, and the pipes' valves are controlled by some temperature sensors located in each boiler. A microcontroller has input ports to receive the bits for the physical parameters, the timer to interrupt at set intervals and inputs, the user will be able to get information about the current temperature in any boiler by simply sending a boiler identification number by using GSM phones. When the temperature inside any boiler reaches a maximum presented value, the system will send an SMS to the user informing that the maximum temperature has been reached.

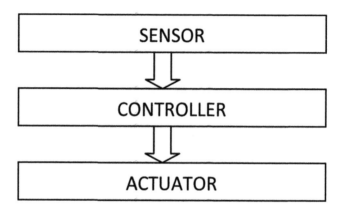

FIGURE 26.4   Diagram of a process controller.

26.3.4   BIOMEDICAL INDUSTRIES

The microcontroller performs various functions, such as processing data from biosensors, storing measurements, and analyzing results in different medical fields. The increasing use of microcontrollers and associated software in both implanted and external medical devices poses special analytical challenges, such as sphygmomanometer, ECG, pacemaker, EEG, pulse oximeter, and oxygen concentrator. One example of such challenges is given in Figure 26.5.[7]

An oxygen concentrator produces a supply of air with increased oxygen content. It can be used to replace liquid oxygen or pressurized oxygen tanks for people who require oxygen-enriched air. Oxygen concentrators work by removing the nitrogen which normally accounts for approximately 78% of the volume of ambient air. The compressor that moves air into the oxygen concentrator and generates the pressure in the sieve beds is driven by an electric motor, making efficient motor control an important part of oxygen concentrator design. Microchip's high performance 16-bit dsPIC30F family of digital signal controllers offers powerful dedicated peripherals to simplify control various types of motors.

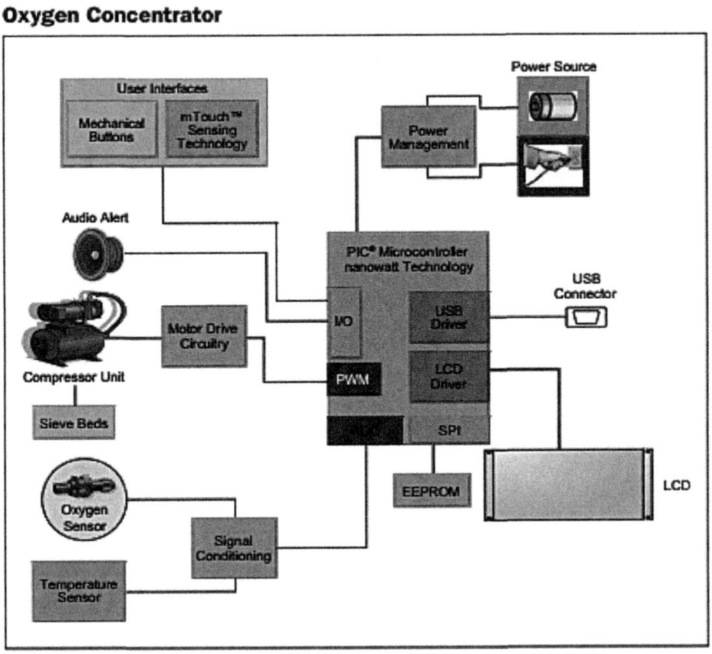

Oxygen Concentrator

FIGURE 26.5   Oxygen concentrator.

26.3.5   IMAGING APPLICATIONS AND SECURITY SYSTEMS

Image processing means frames captured from digital cameras and computations made using pixel values that can be processed through by using microcontroller. One example of image processing with microcontroller is given in Figure 26.6.[2]

Figure 26.6 is a block diagram of a camera interface and object-tracking system. As you can see, the camera is controlled via some of the microcontroller's general purpose I/O pins. The analog output of the camera is attached to the external A/D converter. The servos are connected to two more pins of the microcontroller, and the RS-232 converter conditions the universal asynchronous receiver–transmitter (UART's) signals for connection to the outside world.

Image processing plays a vital role in security systems. One example of security system using image processing with microcontroller is given in Figure 26.7. The block diagram of security system using image processing, touch screen, and verification software is shown. It consists of power supply section, keyboard, verification software, ATMEGA 16 microcontroller, MAX232, touch screen, object, LCD display, and DC motor. Touch screen is used for first-step identification. Keypad is used to enter the code and to answer the security question; LCD displays the entered password or answer of details asked. We can use this system in bank for protecting the locker, and so on. It will provide high-level securities.[10]

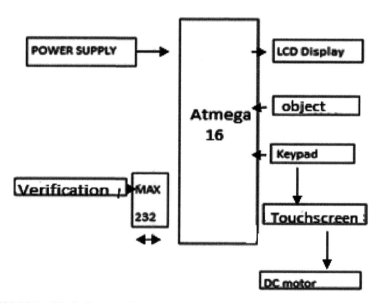

FIGURE 26.7   Block diagram of a security system.

26.4 CONCLUSION

Nonprogrammable controllers would require complex or costly implementations for getting the result but a microcontroller program can be easily modified to different needs of a product line, allowing upgrade in performance with normal redesign, like automobile manufacturing uses microcontroller for engine management system for optimal control of emission for any operating condition. So, microcontrollers evolve in our daily life without any interruption. These include cellular telephones, DVD video system, HDTV broadcasts, etc.

KEYWORDS

- **microcontroller**
- **RISC**
- **CISC**
- **microprocessor**
- **application specific processor**

REFERENCES

1. https://en.wikipedia.org/wiki/Microcontroller.
2. http://www.atmel.com/images/issue4_pg39_43_robotics.pdf.
3. www.ijetae.com (ISO 9001:2008 Certified Journal **2013,** *3* (4)).
4. www.ijarcce.com/.../71-o-karuppiah%20vel-Embedded%20System%20Based%20Ind.
5. *Applications of Microcontrollers*, 2012; Online Material Downloadable at http://www. medgasexperts.com/pdf/Amico/ alarm_OM.pdf.
6. http ://www.atmel.com/applications/industrialautomation/default.aspx.
7. ww1.microchip.com/devicedoc/01062b.
8. www.industry.siemens.com.
9. en.wikipedia.org.
10. http://www.krishisanskriti.org/vol_image/03Jul201503075815.pdf.
11. https://www.pantechsolutions.net/project-kits/mobile-communication-with-microcontroller-8051.
12. *The Uses of Microprocessors*, 2012; an Online Material Downloadable at http://web. engr.oregonstate.edu/~qassim/index_files/Final_ECE570_ASP_2012_Project_Report. pdf.

CHAPTER 27

DESIGN AND PERFORMANCE ANALYSIS OF VARIOUS ADDERS FOR AN ACCUMULATION UNIT OF RRC FILTER

KANAPARTHI REVATHI* and KOTIPALLI PUSHPA

Department of ECE, Shri Vishnu Engineering College for Women, Bhimavaram, Andhra Pradesh, India

Corresponding author. E-mail: revathikanaparthi@gmail.com

ABSTRACT

Adders are the important part of the digital signal processing applications and are digital components most widely used in the digital integrated circuits. With the advances in technology, researchers are trying to design adders with either high speed, low power consumption, less area, or the combination of them. Each adder generates a carry value that has to be transmitted through the design within a series of adders. This generates the critical path delay of the circuit. The latency in the circuit can be reduced by decreasing the number of stages the carry has to be propagated. This chapter describes the analysis of speed, power, and delay of two different types of adder like ripple carry adder and carry select adder for accumulation unit of root raised cosine filter. These two adders are synthesized and simulated using Xilinx ISE 12.2 tool for Spartan 3E family device and simulation results as well as synthesis reports are presented.

27.1 INTRODUCTION

The finite impulse response (FIR) filters are widely used in the mobile communication systems for pulse shaping, channel equalization, and matched filtering due to their properties of linear phase and stability. In a signal processing, filter is used for removing unwanted or feature from the signal. Based on our requirement, different types of filters are used. Root raised cosine (RRC) filters are mostly used in wireless communication system because they have high intersymbol interference rejection ratio than the available other pulse-shaping filters.

In any filter, two major operations are required, they are multiplication and addition. The performance of the filter depends on the accumulation unit of the filter. Here, two different types of the adders are designed to analyze the accumulation unit of RRC filter. In general, addition operation contains two numbers which are added and carry will be produced.[1] The result of the addition process will be sum value and carry value. Half adder and full adder (FA) are basic building blocks that are used to design all complex adder architectures. In this chapter, this section deals with the introduction about filters and adders. In Section 27.2, the features of ripple carry adder (RCA) and carry select adder (CSLA) are discussed. Section 27.3 deals with the introduction about RRC filter. Finally, Sections 27.4 and 27.5 deal with the simulation, synthesis, and comparison results of adders and RRC filter.

27.2   ARCHITECTURE OF ADDERS

The design and features of two different types of adders such as RCA and CSLA[8] are mentioned below. The each adder is named based on the propagation of carry between the stages of the architecture.

27.2.1   RIPPLE CARRY ADDER

FA is a basic adder block in the RCA and it works on basic addition principle.[1] One FA is used for adding 2 bits along with carry bit. The carry of the one full adder is given to the input of the next full adder and so on as shown in the Figure 27.1. Among the entire adders, RCA is slowest but it occupies less area. Connecting the N FAs generates N bit RCA. The latency of the RCA depends upon the number of bits; if the number of bits is more, the delay of the adder also increases.

Critical path is used for calculating the latency of the RCA. RCA block diagram is shown in Figure 27.1. The disadvantage of RCA is overcome by introducing the CSLA architecture in the next section.

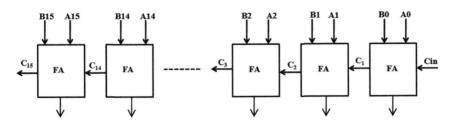

FIGURE 27.1 Architecture of 16 bit ripple carry adder (RCA).

27.2.2 CARRY SELECT ADDER

CSLA is one of the best and fastest adders compared to the RCA and it is also used to perform the fast arithmetic operations in many data processors.[2] CSLA performs the two independent addition operations in parallel using dual RCAs. CSLA structure gives independent outputs sum and carry, that is, $C_{in} = 1$ and $C_{in} = 0$ are done parallelly. Based on C_{in}, the carry is selected by the set of multiplexers to be transmitted to next stage. Further, depending upon the carry input, the sum will be selected. Hence, the latency is minimized. However, the architecture complexity is increased due to number of multiplexers.[4] The structure of CSLA[5] is illustrated in Figure 27.2.

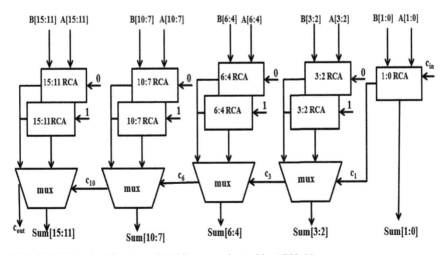

FIGURE 27.2 Architecture of 16 bit carry select adder (CSLA).

27.3   ARCHITECTURE OF RRC FILTER

The filters introduced in mobile communication systems are implemented with low power consumption and high speed. Recently, with the advent of software defined radio concept, finite impulse response filter research has been focused on reconfigurable realizations. These types of reconfigurable filters are used in multiple standards. Reconfigurable RRC filter[3] is important one and architecture of the RRC filter is shown in Figure 27.3. The reconfigurable RRC filter architecture consists of major blocks; they are data generator (DG), a coefficient generator (CG), a coefficient selector (CS), and accumulation unit block. The input signal is given to the DG and it is sampled based on the interpolation selection value. The CG block carry out the multiplication operation between the inputs and filter coefficients. The inputs of the CS are taken from the CG block and it is used to send the correct data to the accumulation unit based on the interpolation selection value. Finally, the accumulation unit is used for summing all outputs of the CSs, and RCC filter output was generated. The above discussed adders are used in the accumulation unit RRC filter for good performance with respect to delay and power.

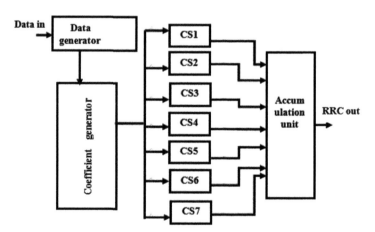

FIGURE 27.3   Architecture of RRC filter with accumulation unit.

27.4   RESULTS AND DISCUSSIONS

Two adders discussed above are programmed by using hardware description language. Synthesis and simulation results are performed by using Xilinx

ISE 12.2 for Spartan 3E family device. In simulation results, technology view designates top block which indicates the set of inputs and outputs. Register transfer logic (RTL) view shows internal block architecture along with the linkage between input and output terminals. Simulation result is produced by writing test bench program for the design. Test bench program has the set of input test vectors that are applied to design.

Simulation result of the RCA is shown in Figure 27.4. Here, a, b, and c_{in} are the inputs and sum and c_{out} are the outputs of the RCA. The RTL view diagram of the RCA is shown in Figure 27.5. Simulation result of the CSLA is illustrated in Figure 27.6. The addition operation is performed between a, b, and c_{in}. The RTL view of the CSLA is shown in Figure 27.7.

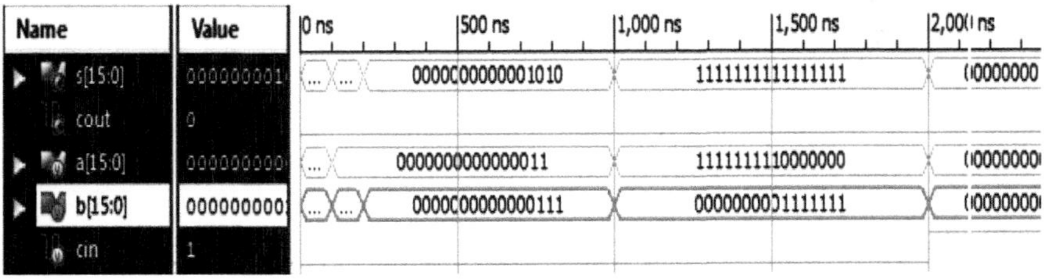

Name	Value	0 ns	500 ns	1,000 ns	1,500 ns	2,000 ns
s[15:0]	000000001	... ...	0000000000001010		1111111111111111	(10000000
cout	0					
a[15:0]	000000000	...	0000000000000011		1111111110000000	(10000000
b[15:0]	000000000	... ...	0000000000000111		0000000001111111	(10000000
cin	1					

FIGURE 27.4 Simulation result of RCA.

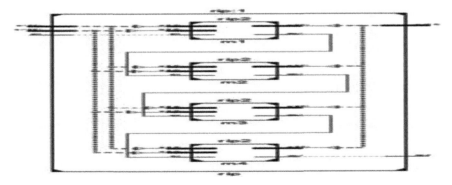

FIGURE 27.5 RTL view of RCA.

27.5 COMPARISON RESULTS

The CSLA architecture achieves less delay and low power when compared to the RCA architecture results. Comparison results of the two adders are shown in Table 27.1. The delay of the RCA is 28.74 ns and power is 0.161 W. The delay of the CSLA is 25.24 ns and power is 0.159 W. The CSLA has less delay and power when compared to the RCA but it occupies more area than the RCA. We can analyze both adders individually and these are used in the accumulation unit of the RRC filter for better performance. The CSLA-used RRC filter has less delay and low power. The RRC filter with CSLA achieves low delay and power compared to the RRC filter with RCA. RRC filter comparison results are mentioned in Table 27.2.

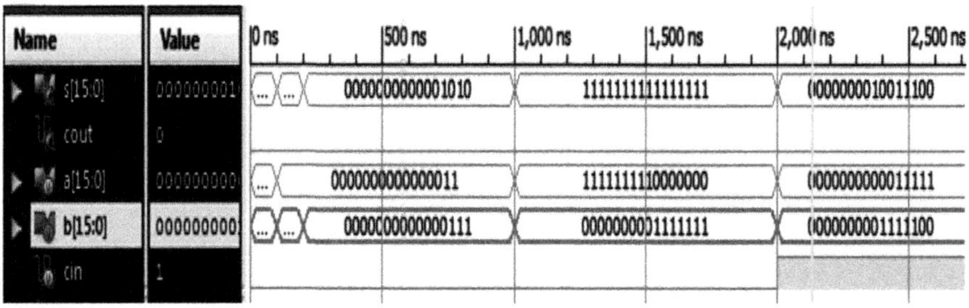

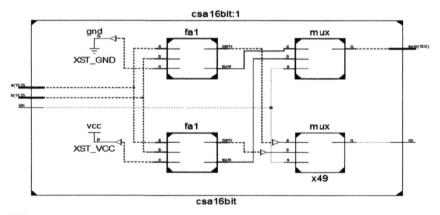

FIGURE 27.6   Simulation result of CSLA.

FIGURE 27.7   RTL view of CSLA.

TABLE 27.1   Adders Comparison Result.

Type of adder	No. of slices	Delay (ns)	Power (W)
RCA	18	28.74	0.161
CSLA	31	25.24	0.159

TABLE 27.2 Adders Used in RRC Filter Comparison Result.

Type of adder	No. of slices	Delay (ns)	Power (W)
Existing method	460	24.52	0.093
Proposed method	594	23.31	0.082

27.6 CONCLUSION

We propose accumulation unit of RRC filter architecture with CSLA which produces the filter output. The proposed architecture gives less delay and low power when compared with the existing RRC filter architecture with RCA. Finally, we can achieve reduced power, delay, and improved performance through this architecture.

KEYWORDS

- carry select adder
- ripple carry adder
- RRC filter
- finite impulse response
- test bench program
- hardware description language
- Xilinx ISE 12.2 for Spartan 3E

REFERENCES

1. Uma, R.; Vijayan, V.; Mohanapriya, M.; Paul, S. Area, Delay and Power Comparison of Adder Topologies. *Int. J. VLSI Des. Commun. Syst.* **2012,** *3* (1), 153–168.
2. Devi, P.; Girdher, A.; Singh, B. Improved Carry Select Adder with Reduced Area and Low Power Consumption. *Int. J. Comput. Appl.* **2010,** *3* (4), 14–18.
3. Nekoei, F.; Kavian, Y. S.; Strobel, O. In *Some Schemes of Realization Digital FIR Filters on FPGA for Communication Applications. 20th International Crimean Conference o Microwave and Telecommunication Technology (CriMiCo), 2010,* Sevastopol, Ukraine, Sept. 13–17, 2010.
4. Senthilkumar, A. VLSI Implementation of an Efficient Carry Select Adder Architecture. *Int. J. Adv. Res. Sci. Eng.* **2013,** *2* (4), 88–93.

5. Singh, S.; Kumar, D. Design of Area and Power Efficient Modified Carry Select Adder. *Int. J. Comput. Appl.* **2011,** *33* (3), 14–18.

6. Jhansi, N.; Jaswanth, B. R. B. Design and Analysis of High Performance FIR Filter Using MAC Unit. *Int. J. Adv. Res. Comput. Commun. Eng.* **2014,** *3* (11), 8626–8629.

7. Gupta, V.; Mohapatra, D.; Raghunathan, A.; Roy, K. Low-power Digital Signal Processing Using Approximate Adders. *IEEE Trans. Comput.-Aided Des. Integr. Circuits Syst.* **2013,** *32* (1), 124–137.

8. Parmar, S.; Singh, K. P. In *Design of High Speed Hybrid Carry Select Adder.* Proceedings of IEEE International Conference on Advance Computing, Nanjing, China, 2012; pp 1656–1663.

PART V

Intelligent Control and Signal Processing Systems

CHAPTER 28

A NEW HYBRID DE–TLBO OPTIMIZATION ALGORITHM FOR CONTROLLER DESIGN AND GLOBAL OPTIMIZATION

PRATEEK DHANUKA[1*], VAIBHAV SINGH RAJPUT[2], BODA BHASKER[3], and RAVI KUMAR JATOTH[4]

[1]*Department of Electronics and Communication Engineering, National Institute of Technology, Andhra Pradesh, India*

[2]*Department of Materials and Metallurgical Engineering, National Institute of Technology, Andhra Pradesh, India*

[3]*Department of Electrical and Computer Engineering, Wollega University, Nekemte, Ethiopia*

[4]*Department of Electronics and Communication Engineering, National Institute of Technology, Warangal, Telangana, India*

Corresponding author. E-mail: prateek.dhnauka03@gmail.com

ABSTRACT

In this chapter, a new hybrid evolutionary algorithm is proposed and applied to a benchmark problem. The benchmark problem is a position servomechanism system. Position servomechanisms are very important in various fields of engineering such as robotics, electrical machines, etc. However, designing a controller can be a tedious task. Many algorithms have been applied to accomplish this task. Recently, a very powerful metaheuristic algorithm was proposed. This algorithm ended up being excellent in finding global optima for low-dimensional problems. So, this algorithm found fame quite quickly and has been successfully hybridized with various other algorithms. In this chapter, it was hybridized with another evolutionary algorithm differential evolution (DE). The hybrid was applied to the benchmark problem and results were compared. From the simulation and results, it can be observed that the hybrid DE–TLBO-based controller outperformed other existing algorithms.

28.1 INTRODUCTION

In the past few years, there has been a growing interest in evolutionary computing, which has led to the development of many optimization algorithms, each one better than the previous one. Many engineering problems can be viewed as optimization problems such as economic dispatch problem, pressure vessel design, DC motor position control, VLSI design, etc. Due to this, the engineering community has shown a significant interest in this field. One such algorithm is differential evolution proposed by Storn and Price in 1995.[1] Originally, Price and Storn had proposed a single strategy for the compact and simple algorithm, which they later expanded to 10.[2] It has been praised for its simplicity and compact size; however, it has also been criticized for having a slower converging rate than other evolutionary algorithms. This can be rectified by changing the few parameters on which the algorithm relies. However, it should be noted that these parameters do not affect the quality of the optimum value achieved. Recently, a novel and effective algorithm simulating the teaching and learning phases of a classroom was proposed,[3–5] appropriately named teacher learning-based optimization (TLBO); it is highly praised for its simple concept and high efficiency. Due to this reason, TLBO has become a very attractive algorithm and has been applied to many real-world engineering applications. Hybridization of algorithms has become common practice in the past few years as not only does this increase the converging capabilities of both the algorithms, but also it combines the characteristics of the two to give more desirable characteristics. In this case, we combine two relatively young algorithms, namely differential evolution and TLBO. Differential evolution is known for its ability to converge at the global minima no matter what initial parameters are assigned to it. This process, however, can take large number of iterations in computationally expensive functions. On the other hand, TLBO is an efficient algorithm which converges to the local minima rather quickly in lower dimension problems. Hybridizing the two algorithms can yield better results in less computational time. In this chapter, the hybrid algorithm is tested upon DC motor with elastic shaft, a benchmark system from Ref. [6]. The system is a position servomechanism, requiring a controller to control the final position of load shaft when external disturbances are observed. In this chapter, a proportional-integral-derivative (PID) controller is tuned using the evolutionary algorithms. A PID controller has three major parameters on which the controller depends, K_p, K_i, and K_d.

$$u(t) = K_p \cdot e(t) + K_i \int_0^t e(t) + K_d \frac{de(t)}{dt}$$

The results are compared with the results obtained from the individual algorithms.

28.2   PROBLEM FORMULATION

PID control is the most common control algorithm used in industry and has been universally accepted in industrial control. Their popularity can be accredited to their robust performance in wide range of fields. As the name suggests, PID algorithm consists of three basic coefficients, proportional, integral, and derivative gains which yield different results based on their values. Finding the optimum values of these coefficients is known as PID tuning. This however can be very exhausting and time-consuming as most tuning methods involve trial and error to find one or more of the parameters. This is much clearer from Figures 28.1 to 28.3.

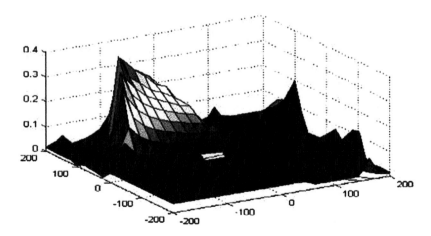

FIGURE 28.1   Error when K_p and K_d are altered keeping K_i constant.

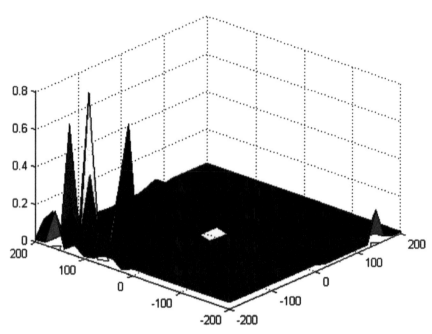

FIGURE 28.2   Error when K_i and K_d are altered keeping K_p constant.

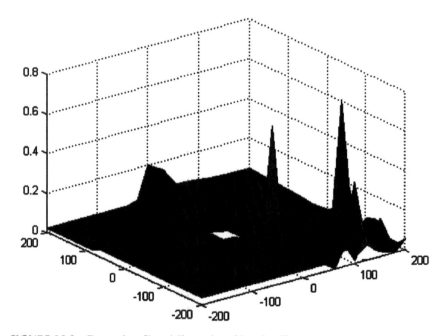

FIGURE 28.3   Error when K_p and K_i are altered keeping K_d constant.

28.3   MATHEMATICAL MODELING

The benchmark position servomechanism system is given below. To make this chapter self-containing, the whole system is restated along with the derivation of its state-space equation. As depicted in Figure 28.4, the system consists of a DC motor, a gearbox, an elastic shaft, and a load. Now, we derive the equations of the system to obtain a model suitable for control tasks. For this aim, let us consider each physical component of the system.

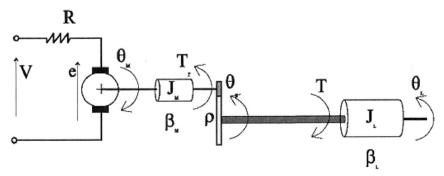

FIGURE 28.4   The benchmark position servomechanism system.

28.3.1   DC MOTOR—ARMATURE CONTROL

$$V = iR + e$$

where V is the applied armature voltage, i is the armature current, R is the resistance, and e is the back EMF.

$$e = K_b \, \omega_m$$

where $\omega_m = \Theta_m$ is the angular velocity of motor shaft and K_b is the constant.

$$\Phi = K_f i_f$$

where Φ is the air gap flux, i_f is the field current, and K_f is the constant.

$$T_m = K_1 \Phi$$

where T_m is the torque developed by the motor and K_1 is the constant.

$$T_m \omega_m = ei$$

And therefore, $K_1 K j_f = K_b = K_T$, K_T is the motor constant.

$$J_m \dot{\omega}_m = T_m - \beta_m \omega_m - T_r$$

where J_m is the equivalent moment of inertia of motor, ω_m is the equivalent vicious friction coefficient of motor, and T_r is the other torques.

In summary,

$$V = iR + K_T \omega_m$$

$$J_{mm} = T_m - \beta_m \omega_m - T_r$$

28.3.2 GEAR BOX

Consider the gear box given below (Fig. 28.5):

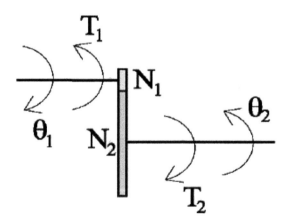

FIGURE 28.5 Gear box.

$$\theta_1 r_1 = \theta_2 r_2$$

where r_1 and r_2 are the wheel radii.

$$\frac{\theta_1}{\theta_2} = \frac{r_2}{r_1} = \frac{N_2}{N_1} = \rho$$

By differentiation,

$$\omega_1 = \rho \omega_2$$

Power transmission (no loss)

$$\frac{T_1}{T_2} = \frac{1}{\rho}$$

By referring to Figure 28.5, we have $\theta_1 = \theta_m$, $\theta_2 = \theta_s$, $T_1 = -T_r$, $T_2 = T$.

$$\theta_s = \frac{1}{\rho}\theta_m$$

$$T_r = -\frac{1}{\rho}T$$

where θ_m is the angular displacement of motor shaft, θ_s is the angular displacement of load-side gear, and T is the torque acting on the load.

28.3.3 ELASTIC SHAFT

The shaft has finite torsional rigidity K_θ:

$$T = K_\theta\left(\theta_L - \theta_s\right)$$

where θ_L is the angular displacement of load.

28.3.4 LOAD DYNAMICS

$$J_L\dot{\omega}_L = -\beta_L\omega_L - T$$

28.3.5 DIFFERENTIAL EQUATIONS OF THE SYSTEM

By collecting the previous equations, we can write

$$\dot{\omega}_L = -\frac{K_\theta}{J_L}\left(\theta_L - \frac{\theta_M}{\rho}\right) - \frac{\beta_L}{J_L}\omega_L$$

$$\dot{\omega}_M = \frac{K_T}{J_M}\left(\frac{V - K_T\omega_M}{R}\right) - \frac{\beta_M\omega_M}{J_M} + \frac{K_\theta}{\rho J_M}\left(\theta_L - \frac{\theta_M}{\rho}\right)$$

28.3.6   STATE-SPACE MODEL

By setting $x_p = \left[\theta_L, \omega_L, \theta_M, \omega_M\right]$, the system can be described by the following state-space form,

$$\dot{x}_p = \begin{bmatrix} 0 & 1 & 0 & 0 \\ -\dfrac{K_\theta}{J_L} & -\dfrac{\beta_L}{J_L} & \dfrac{K_\theta}{\rho J_L} & 0 \\ 0 & 0 & 0 & 1 \\ \dfrac{K_\theta}{\rho J_M} & 0 & -\dfrac{K_\theta}{\rho^2 J_M} & -\dfrac{\beta_M + K_T^2/E}{J_M} \end{bmatrix} x_p + \begin{bmatrix} 0 \\ 0 \\ 0 \\ \dfrac{K_T}{R J_M} \end{bmatrix} V$$

$$\begin{bmatrix} \theta_L \\ \dot{\theta}_L \\ T \end{bmatrix} = \begin{bmatrix} 1 & 0 & 0 & 0 \\ 0 & 1 & 0 & 0 \\ K_\theta & 0 & -\dfrac{K_\theta}{\rho} & 0 \end{bmatrix} x_p$$

The designed controller must set the load's angular displacement to the desired value. Since the elastic shaft has finite shear strength, so the torque, T, must stay within the limits, $|T| \leq 78.5$ N m. Also, the applied voltage must stay within the limits, $|V| \leq 220$ V (Table 28.1).

TABLE 28.1   System Parameters.

Symbol	Value (SI unit)	Symbol	Value
K_q	1280.2	ρ	20
K_T	10	B_M	0.1
J_M	0.5	β_L	25
J_L	50 JM	R	20

28.4   DIFFERENTIAL EVOLUTION

Differential evolution (DE) is a powerful optimization algorithm with very few algorithm-specific parameters. DE relies on a randomly initialized population comprising N_p individuals. Each individual $x_i = \{x_{i1}, x_{i2}, x_{i3}, \ldots, x_{in}\}$ is a vector of n dimensions. DE relies on three operators that are repeated for It_{max} number of iterations. The operators are namely mutation, crossover, and selection.

28.4.1 MUTATION

This operation creates a mutant vector u_i by selecting components from a randomly selected vector x_a and the difference of two other randomly generated vectors x_b and x_c. Mathematically,

$$u_i = x_a + \beta \times (x_b - x_c)$$

where $a \neq b \neq c \neq i$. β is a random number used to control the perturbation size of the mutation. Here, x_a is known as base vector.

28.4.2 CROSSOVER

This operator creates a trial vector v_i by crossing over mutant vector and target vector (another randomly generated vector). In other words, trial vector is generated by randomly selecting components from mutant vector (u_i) and trial vector (x_i) using a probability factor (p_{CR}). Mathematically,

$$v_{ij} = \begin{cases} u_{ij}, & \text{if } \text{rand} \leq p_{CR} \text{ or } j = j_0 \\ x_{ij}, & \text{otherwise} \end{cases}$$

The probability factor (p_{CR}) controls the diversity of the population and helps the algorithm to escape from local minima. j_0 is a randomly generated index between $\{1, 2, 3, \ldots, N_p\}$. This guarantees that v_i has at least one component from u_i.

28.4.3 SELECTION

This operator chooses the better offspring among v_i and x_i using their fitness. Mathematically,

$$x_i = \begin{cases} u_i, & \text{if } \text{fitness}(u_i) > \text{fitness}(x_i) \\ x_i, & \text{otherwise} \end{cases}$$

This operator guarantees that each iteration solution is better than the solution obtained in the previous iteration.

28.5   TEACHER LEARNING-BASED OPTIMIZATION

TLBO is a new algorithm by Rao et al.[3] It is inspired by the classroom environment and can be termed as a simulation of modern education. It can be divided into two phases.

28.5.1   TEACHER PHASE

A teacher can be considered to be the most educated individual in the society. Hence, the student with the highest marks acts as a teacher during the teacher phase. The teacher tries to enhance the mean of the class to her level. This, however, depends on the learning capability of the class. This is formulated as

$$x_{i_{temp}} = x_i + rand \times (Teacher - TF \times Mean)$$

where TF = ceil (0.5 + rand) is the teaching factor and mean is the mean of the class. The new solution, $x_{i_{temp}}$, is accepted only if it is better than the previous solution, that is

$$x_i = \begin{cases} x_{i_{temp}}, & f(x_{i_{temp}}) > f(x_i) \\ x_i, & \text{otherwise} \end{cases}$$

28.5.2   LEARNER PHASE

Teaching is not the only education students receive, they also learn by interacting among each other. This is simulated in the learner phase. In each iteration, two students x_m and x_n interact among each other, with the smarter one enhancing the others' marks. It can be formulated as

$$x_{m_{temp}} = \begin{cases} x_m + rand \times (x_m - x_n), & f(x_m) > f(x_n) \\ x_m + rand \times (x_n - x_m), & f(x_n) > f(x_m) \end{cases}$$

The temporary solution is accepted only if it is better than the previous solution, that is

$$x_m = \begin{cases} x_{m_{temp}}, & f(x_{m_{temp}}) > f(x_m) \\ x_m, & \text{otherwise} \end{cases}$$

28.6   HYBRID DE–TLBO

This section discusses the rationale of the proposed hybrid algorithm. Both DE and TLBO are population-based algorithms with individuals being considered as vectors in DE and learners in TLBO. Initially, DE is initialized with random vectors. TLBO is used as an intermediate algorithm to improve the worst results obtained using DE between generations. TLBO is initialized with lower half of the DE population and random particles between the best and worst results. The newly initialized learners undergo a certain number of TLBO iterations before being sorted and fed back to the DE iterations. The process might be better understood by the flowchart given in Figure 28.6.

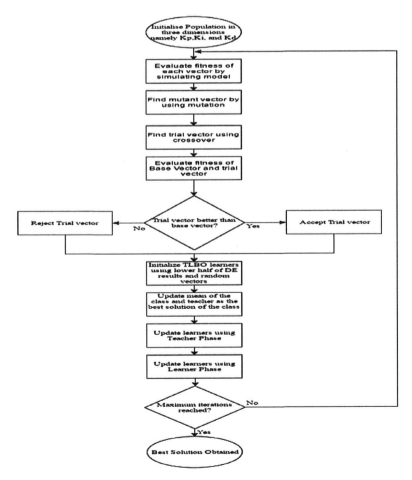

FIGURE 28.6   Flowchart of hybrid DE–TLBO.

28.7 BENCHMARK FUNCTIONS

The proposed algorithm was applied to five benchmark functions from
Ref. [4. The results of TLBO and iTLBO are collected from Ref. [4]
(Table 28.2).

TABLE 28.2   Comparison of Hybrid DE–TLBO Against Other Algorithms.

Algorithm	Sphere	Rosenbrock	Ackley	Griewank	Schwefels
TLBO	0 ± 0	1.72 ± 0.0662	$3.55e - 15 \pm$ $8.32e - 31$	0 ± 0	$2.94e + 02 \pm$ $2.68e + 02$
iTLBO	0 ± 0	2 ± 0.142	$1.42e - 15 \pm$ $1.83e - 15$	0 ± 0	$1.10e + 02 \pm$ $1.06e + 02$
Hybrid DE–TLBO	0 ± 0	0 ± 0	$1.42e - 15 \pm$ $1.83e - 15$	0 ± 0	$7.91 \pm 2.95e$ $+ 01$

28.8 IMPLEMENTATION OF PROPOSED ALGORITHM TO PID TUNING

PID tuning can be viewed as a three-dimensional problem with the dimensions being the values of K_p, K_i, and K_d. The proposed hybrid DE–TLBO can
be used to find the global optimum, which, in this problem, is the value of
the PID gains for which minimum error is obtained.

28.8.1 INITIALIZING OF THE SOLUTIONS

The population P consists of N_p particles each having three elements. The
first element represents the value of K_p, the second element represents the
value of K_i, and the third element represents the value of K_d. Any particle x_i
can be represented as

$$x_i = \begin{bmatrix} K_p, & K_i, & K_d \end{bmatrix}$$

28.8.2 APPLYING THE ALGORITHM

The values of K_p, K_i, and K_d have to be used to simulate a Simulink model
each time to calculate the fitness of any particle. There are various methods

of calculating the fitness such as integral absolute error, integral square error, integral time square error, integral time absolute error (ITAE), etc. In this chapter, ITAE was used to calculate the fitness of the particles. Using the values obtained from the chosen method, the Simulink model can be optimized, in this case, a DC motor with an elastic shaft.

28.9   RESULTS

The results of K_p, K_i, and K_d obtained using hybrid DE–TLBO were used and the rise time, settling time, maximum overshoot, and steady state error were calculated. These values were compared with those obtained from DE and TLBO.

The results are tabulated in Table 28.3.

For an optimally searched PID controller using DE,

$$G_c(s) = 109.9551 + \frac{0.02758}{s} - 11.7257s$$

For an optimally searched PID controller using TLBO,

$$G_c(s) = 105.0846 - \frac{0.06812}{s} - 120.7683s$$

For an optimally searched PID controller using hybrid DE–TLBO,

$$G_c(s) = 173.6313 - \frac{0.0001593}{s} - 75.7108s$$

TABLE 28.3   Comparison of Hybrid DE–TLBO Against Other Algorithms.

Algorithm	Rise time (s)	Settling time (s)	Steady state error (%)	Maximum overshoot (%)
TLBO	0.75	1.64	−0.050255	0.067879
DE	0.75	1.57	0.032103	0.024442
Hybrid DE–TLBO	0.63	2.14	0.076469	7.9783e − 05

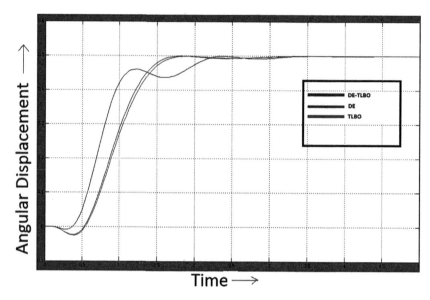

FIGURE 28.7 **(See color insert.)** Comparison of different algorithms.

28.10 CONCLUSIONS

As the results indicate, hybrid DE–TLBO is providing much better results than the base algorithms. The required output is 0.5 units. The rise time obtained is 0.61 s, the settling time is 2.14 s, the maximum overshoot is 0.076469%, and the steady state error is 7.9783×10^{-5}%. The rise time, steady state error, and overshoot are better than the other controllers. In case of settling time, the controller may be seen as worse; however, the major upside, as evident from Figure 28.7, is the undershoot obtained in the controller designed using DE–TLBO.

KEYWORDS

- evolutionary computing
- optimization algorithms
- engineering problems
- global optimization

REFERENCES

1. Price, K.; Storn, R. Differential Evolution—A Simple and Efficient Heuristic for Global Optimization over Continuous Spaces. *J. Global Optimizat.* **1997,** *11*, 341–359.
2. Price, K. An Introduction to DE. In *New Ideas in Optimization*; Corne, D., Marco, D., Glover, F., Eds.; McGraw-Hill: London, UK, 1999; pp 78–108.
3. Rao, R. V.; Savsani, V. J.; Vakharia, D. P. Teacher Learning Based Optimization: A Novel Method for Constrained Mechanical Design Optimization Problems. *Comput.-Aided Des.* **2011,** *43*, 303–315.
4. Rao, R. V.; Patel, V. An Elitist Teaching Learning Based Optimization Algorithm for Solving Complex Constrained Optimization Problems. *Int. J. Ind. Eng. Comput.* **2012,** *10*, 535–560.
5. Rao, R. V.; Savsani, V. J.; Vakharia, D. P. Teacher-learning-based Optimization: An Optimization Method for Continuous Non-linear Large Scale Problems. *Inf. Sci.* **2012,** *183*, 1–15.
6. Mathworks Inc. *Position Servomechanism System* [Online]. http://in.mathworks.com/help/mpc/ug/servomechanism-controller.html.

CHAPTER 29

PERFORMANCE COMPARISON OF PSO ALGORITHMS FOR MOBILE ROBOT PATH PLANNING IN COMPLEX ENVIRONMENTS

RAVI KUMAR JATOTH[1*], K. JAYA SHANKAR REDDY[2], K. KARTHIKEYA YADAV[2], and BODA BHASKER[3]

[1]*Department of Electronics and Communication Engineering, National Institute of Technology, Warangal, Telangana, India*

[2]*Department of Computer Science and Engineering, National Institute of Technology, Andhra Pradesh, India*

[3]*Department of Electrical and Computer Engineering, Wollega University, Nekemte, Ethiopia*

Corresponding author. E-mail: jrk.nitw@gmail.com

ABSTRACT

This chapter compares the performance of different particle swarm optimization (PSO) algorithms when applied to the problem of mobile robot path planning in complex environments. The main focus is on the paths that are feasible for a mobile robot by avoiding obstacles in complex environment. A constrained environment is chosen where a robot is represented as a single point. PSO is one of the best evolutionary algorithms applied for robot path planning. There are many improvised variations of PSO modifying the classical PSO. Four different variations of PSO are applied for mobile robot path planning and the results are compared.

29.1 INTRODUCTION

Path planning is a fundamental problem in mobile robotics. It is the process of generating a feasible path for a mobile robot in such a way that the robot avoids obstacles. The robot path planning is classified as local and global.[1-3] In local path planning, the robot reaches the goal in steps evolving its next best position each time in an unknown or known environment, whereas in global path planning, the robot first reaches the goal and tries different paths to avoid obstacles. Global path planning is also referred as offline path planning and local path planning as real-time path planning. Every path which directs the robot from source to the desired target is a feasible path.[4] Generally, path planning involves two main aims: (1) the path should be feasible and (2) the path should also avoid obstacles. Achieving the above two aims enables the robot path planning. In practical cases, robot path planning is done by detecting the obstacles using image processing both either in known and unknown environments or even in static and dynamic environments. But in optimization techniques, we do not use image processing for detecting obstacles. So it becomes a bit difficult to generate a path in an unknown environment using optimization.[5] In this chapter, we use a known environment in which there are geometrical obstacles.

For robot path planning, we can use suitable evolutionary algorithms out of which particle swarm optimization (PSO)[7] is in our interest. Basic types of PSO algorithms are adaptive PSO, binary PSO, and the modified PSO. All these are discussed below.

29.2 PROBLEM FORMULATION

The problem is stated as follows. The robot is considered as a single point and moves in a closed worked space. The workspace is a two-dimensional environment containing static and geometrical obstacles. The source point and the desired goal point are chosen. The objective is to generate a collision-free path taking the robot from the source point and the goal point. The path is divided into segments connecting points from the source to the goal. The area in the workspace occupied by the segments of the path is the configuration space (C-space). Practically, C-space is the region obtained by sliding the robot along the edges of the obstacles. The complexity of the path planning increases as the number of dimensions of the C-space increases.

The path is made not to go out of the C-space by applying the limits of position and the velocities. The path will be smooth only if the obstacles do not have sharp corners. But in complex cases, there might be sharp cornered obstacles. So, we can imagine them blunt by circumscribing or inscribing a circle of fewer radiuses around the obstacles. The path will avoid the circles which imply that the original obstacles are avoided.

Looking for the shorter path does not mean that the time taken is less; we need a complex algorithm for a complex environment where the time taken to generate the shortest path might be longer.

29.3   METHODOLOGY

Figure 29.1 is a small example that illustrates the robot path planning in a C-space.

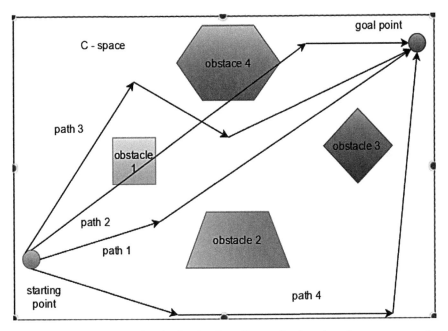

FIGURE 29.1   Example of path planning in a C-space having obstacles, a source, and a goal point.

Path 1: avoids obstacles and optimizes the path; path 2: does not avoid the obstacles; path 3: avoids obstacles but does not optimize the path; path 4: the path that did not meet our conditions.

Obstacles

- Occupied spaces of the world
- Robot should not go into that space

C-space

- Unoccupied spaces within the world
- Where the robot can move
- Also referred as the workspace or the boundaries

Inputs

- Geometry of the robot
- Geometry of the obstacles
- Geometry of the free space
- A starting and the desired goal position

Outputs

A continuous path connecting the source and the goal.

First, some points or locations in the C-space are to be chosen, they are connected with each other to form a path from the source to the goal point and then try to avoid the obstacles. The optimization goals of the path planning are as follows:

- The distance traveled by the robot should be least.
- The path should not run into obstacles.
- The path should be smooth.
- The path should not lead the robot outside the C-space.

29.4   OBJECTIVE FUNCTION

The objective function is the basis by which the path is optimized in the robot path planning. There are different objective functions by which robot path planning[6] is done. One of them is by calculating the length of each particle, that is, the sum of each segment in the particle.

$$d = \sqrt{(x_2 - x_1)^2 + (y_2 - y_1)^2}$$

where the line joining the points (x_1, y_1) and (x_2, y_2) is one of the segments of the path.

Now add all such segments to d.

$D =$ sum (d) (sum of all segments)

Now, D gives the distance of the entire path from source to goal.

This just optimizes the distance traveled by the robot but we also have to avoid the obstacles, for that, we introduced a term foul, that is, if any of the coordinates of the path lies in the boundary of obstacles, and then add this foul to the objective function. Before adding to it, the foul is increased by a factor, K, which is chosen as 100 in this path planning.

Since each obstacle is circumscribed or inscribed by a circle, we can say that the path is foul if any of the coordinates of the path is inside a circle.

If the distance between the point and the center of a circle is less than the radius, then the point is said to be inside the circle.

Foul = 0 (Initially)

For $i = 1$: no of points

$$d = \sqrt{\left(x_i - c_1\right)^2 + \left(y_i - c_2\right)^2}$$

$[(c_1, c_2)$—Center of the circle]

If $d < r$; r—radius of the circle.

Foul = Foul + $(r - d)$

Now the objective function becomes

$D = D \times (K \times$ foul)

Example of a foul path with square obstacle is shown in Figure 29.2.

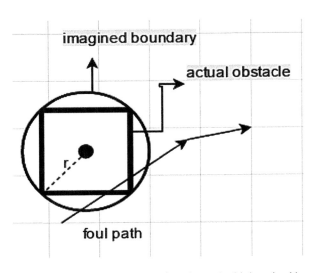

FIGURE 29.2 Example of a path that meets obstacles and with imagined boundary.

So, the objective function's value will be less if the path distance is shorter and there is no foul in the path.

Hence, this is our objective function for robot path planning.

29.5 OPTIMIZATION ALGORITHMS

29.5.1 CLASSICAL PSO

In the PSO, the possible solutions of the objective function have initialized as the particles of the swarm. As an algorithm, the main strength of PSO is its fast convergence. These particles are distributed in the C-space.

In every iteration, the positions and velocities of all the particles are updated.

Velocity update:

$$V_i = w \times V_i + C_1 \times r_1 \times (P_i - X_i) + C_2 \times r_2 \times (G_i - X_i) \qquad (29.1)$$

where C_1, C_2, and w are the coefficients of self-component, social component, and inertial weight, respectively.

r_1, r_2—Random numbers in $[0, 1]$
Position update:

$$P_i = P_i + V_i \qquad (29.2)$$

This replaces the old particle with the updated particle and calculates the function value of these particles. Now update the new particle best and find out the global best among these and store such at each iteration. At the end of iterations, the best function value can be seen in the plot or can be displayed.

29.5.2 ADAPTIVE PSO (PSO–TVIW)

This version is same as classical PSO except that the inertial weight differs at each iteration. As the number of iterations increases, the inertial weight goes on decreasing by using a formula.

$$w = w_{\max} - \left(\frac{w_{\max} - w_{\min}}{\text{maxiter}} \right) \times \text{iter}$$

where maxiter is the total number of iterations and iter is the present iteration.

This can also be done by a damping factor wd whose value varies the performance of the PSO by a larger extent which is shown in the example of robot path planning later in this chapter.

At the end of each iteration, $w = w \times wd$ is to be done. Initially, $w = 1$.

For robot path planning, the values used were

$$w_{max} = 0.9, w_{min} = 0.4 \text{ and } wd = 0.98$$

The adaptive PSO is also known as PSO–TVIW (time-varying initial weight)

29.5.3 BINARY PSO

The binary PSO was proposed by Eberhart and Kennedy to optimize functions even in binary PSO[8]

This is the extension of the adaptive PSO and varies from it in the velocity updating. The adaptive PSO does not take into account the particles' velocities reaching the maximum.[10]

This PSO is only the first type of the binary PSO. In binary PSO, the position of the particles is updated only after the velocities are reflected. This operation is termed as a mutation. This uses a mutation factor "rmu."

While (number of dimensions)

$r =$ rand ();

If r < rmu

$$v_{id} = -v_{id} \tag{29.3}$$

Stop

The mutation factor rmu is chosen as 0.4.

In the case of robot path planning, the binary PSO gives better results in the much complex environment than other versions of PSO.

The pseudo code is as follows:

Start

$G = 0$ (generation index)

Initialize the swarm with some random positions

Evaluate their function values using the objective function

$G = 0$ (generation index)

Update the velocity using (29.1)

Mutate the velocity using (29.3)

Update the position using (29.2)

Evaluate the function value using objective function

Replace with the old particle

Evaluate the global best for each iteration

If satisfied

Stop;

Else

Go to step 6.

29.5.4 MODIFIED PSO

The classical PSO takes into account only the particles' best position and the global best among all the particles. But the modified PSO takes even the particles' worst position and the global worst among all the particles and also includes them in the velocity updating formula of the particles.

$$V_i = w \times V_i + C_1 \times r_1 \times (P_i - X_i) + C_2 \times r_2 \times (P_i - X_i) + C_3 \times r_3 \times (W_i - P_i) \quad (29.4)$$

W_i is the worst function value of the ith particle; C_3 is another acceleration coefficient; and r_3 is the random number in [0, 1].

And at the end of each iteration, both the global best and the worst best have to be found out. Rest all the steps remain same as the classical PSO.[11]

All the above versions seem to be changed just a little bit in terms of the algorithm from the classical version of PSO, but the path planning varied by a larger extent in different environments.

29.6 SIMULATION AND RESULTS

29.6.1 ROBOT PATH PLANNING IN ENVIRONMENT 1

29.6.1.1 CLASSICAL PSO-BASED ROBOT PATH PLANNING

The classical PSO algorithm generated a path that is free from obstacles but has taken many turns, thereby increasing the length of the path. The path is also not so close to the obstacles as shown in Figure 29.3.

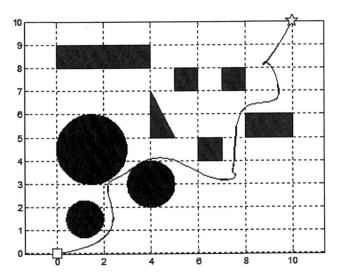

FIGURE 29.3   Path obtained using classical PSO algorithm.

29.6.1.2   ADAPTIVE PSO-BASED ROBOT PATH PLANNING

The adaptive PSO generated a perfect path free from obstacles and also the shortest path but the path is too close to the obstacles as shown in Figure 29.4.

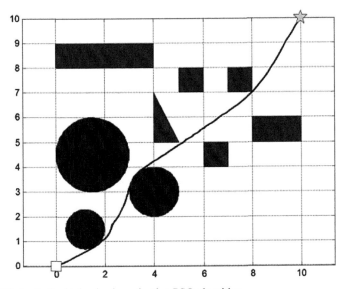

FIGURE 29.4   Path obtained using adaptive PSO algorithm.

29.6.1.3   BINARY PSO-BASED ROBOT PATH PLANNING

This algorithm generated a path that is free from the obstacles but it did not avoid the obstacles in the shortest path. The path is also very close to the obstacles as shown in Figure 29.5.

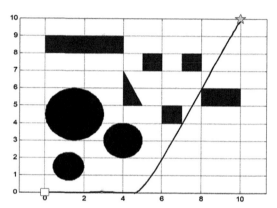

FIGURE 29.5   Path obtained using binary PSO algorithm.

29.6.1.4   MODIFIED PSO-BASED PATH PLANNING

This case is the same as the classical PSO. The path has taken too many turns which is practically not possible for a mobile robot. The path has just tried to avoid the obstacles but did not optimize the path traveled as shown in Figure 29.6.

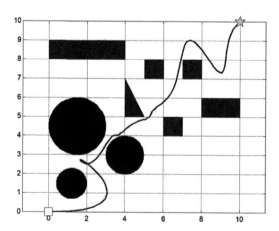

FIGURE 29.6   Path obtained using modified PSO algorithm.

TABLE 29.1   Comparison of the Above Algorithms in Environment 1.

Type of PSO	Path length
PSO classical	19.3366
PSO–TVIW	14.5991
BPSO	16.0705
Modified PSO	21.2

29.6.2   ROBOT PATH PLANNING IN ENVIRONMENT 2

This environment is a bit much complex than the environment 1 as this path has to follow a zigzag pattern to reach the goal avoiding obstacles.

29.6.2.1   ADAPTIVE PSO-BASED ROBOT PATH PLANNING

The classical PSO and the modified PSO had resulted in generating the same path as that of the classical PSO as shown in Figure 29.7.

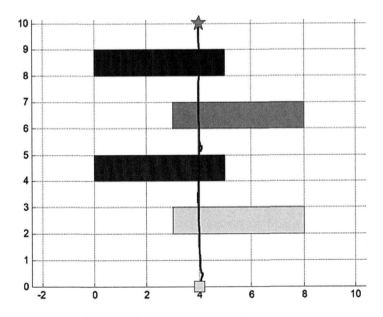

FIGURE 29.7   Path obtained using the adaptive PSO algorithm.

29.6.2.2 BINARY PSO-BASED PATH PLANNING

Binary PSO[9] was avoiding obstacles and reached the goal position. It has followed a zigzag pattern as shown in Figure 29.8.

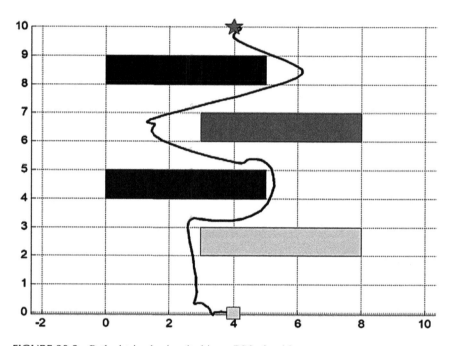

FIGURE 29.8   Path obtained using the binary PSO algorithm.

29.7 CONCLUSION

Four different versions of PSO with the same objective function to optimize the path were applied for a single-point mobile robot path planning. These include the classical PSO, binary PSO, adaptive PSO, and the modified PSO. All of these tried to avoid the obstacles but only the adaptive PSO optimized the path traveled to a large extent in the environment 1. In another much complex environment 2, the binary PSO gave the best path that is feasible for a robot. This concludes that the type of algorithm to be used depends on the geometry of the obstacles and also the environment and the observations are tabulated in Table 29.2.

TABLE 29.2 Overall Comparison of Different PSO Algorithms in Environment 1 and Environment 2.

PSO	Environment 1	Environment 2
Classical	Yes (but not optimize the path length)	No
Adaptive	Yes	No
Binary	Yes (too close to the obstacles)	Yes
Modified	Yes (too many turns in the path)	No

KEYWORDS

- **mobile robot**
- **path planning**
- **classical PSO**
- **binary PSO**
- **adaptive PSO**
- **modified PSO**
- **complex environments**

REFERENCES

1. Regele, R.; Levi, P. In *Cooperative Multi-robot Path Planning by Heuristic Priority Adjustment*. Proceedings of the IEEE/RSJ International Conference on Intelligent Robots and Systems, 2006.
2. Sedighi, K. H.; Ashenayi, K.; Manikas, T. W.; Wainwright, R. L.; Tai, H. In *Autonomous Local Path Planning for a Mobile Robot Using a Genetic Algorithm*. Proceedings of the IEEE International Conference on Robotics and Automation, 2004; pp 1338–1345.
3. Saska, M.; Macas, M.; Preucil, L.; Lhotska, L. In *Robot Path Planning using Particle Swarm Optimization of Ferguson Splines*. Proceedings of 11th IEEE International Conference on Emerging Technologies and Factory Automation, 2006; pp 833–839.
4. LaValle, S. M.; Hutchinson, S. A. In *Optimal Motion Planning for Multiple Robots Having Independent Goals*. IEEE Transaction on Robotics and Automation, 1998; pp 912–925.
5. Hocaoglu, C.; Sanderson, A. C. Planning Multiple Paths with Evolutionary Speciation. *IEEE Trans. Evol. Comput.* **2001,** 5 (3), 169–191.
6. Behnke, S. In *Local Multiresolution Path Planning*. 7th RoboCup International Symposium, Padua, Italy, 2004; Springer, 2004.
7. Eberhart; Shi, Y. In *Particle Swarm Optimization: Developments, Applications, and Resources*. Proceedings of the 2001 Congress on Evolutionary Computation, 2001.

8. Glavaski, D.; Volf, M.; Bonkovic, M. Mobile Robot Path Planning Using Exact Cell Decomposition and Potential Field Methods. *WSEAS Trans. Circuits Syst.* 2009.

9. Eberhart, R. C. In *A Discrete Binary Version of the Particle Swarm Algorithm*. Proceedings of Conference Systems Man Cybernetics, Piscataway, NJ, 1997; pp 4104–4108.

10. Lee, S.; Park, H.; Jeon, M. Binary Particle Swarm Optimization with Bit Change Mutation. *IEICE Trans. Fundament., Electron. Commun. Comput. Sci. E* **2007,** *90A* (10), 2253–2256.

11. Yang, B.; Zhang, Q. Frame Sizing and Topological Optimization Using a Modified Particle Swarm Algorithm. *Second WRI Global Congress, Intelligent Systems (GCIS),* 2010.

CHAPTER 30

IMPLEMENTATION OF AN EFFICIENT AND FULLY AUTOMATED MAGNETIC RESONANCE IMAGE SEGMENTATION THROUGH MACHINE LEARNING

K. V. SRIDHAR* and I. HEMANTH KUMAR

Department of ECE, NIT, Warangal, Telangana, India

Corresponding author. E-mail: sridhar@nitw.ac.in

ABSTRACT

Magnetic resonance (MR) image segmentation is a crucial step in analyzing and processing the medical images. But the noise in the process of MR image acquisition such as intrinsic soft tissue variation, partial volume, overlapped intensities of the tissues and nonuniformity in magnetic field applied in MRI makes medical image segmentation as a challenging task. Generally, the image segmentation results are used as input to further processing algorithms for feature classification. But these segmentation algorithms are computationally intensive and take much more computation time. In case of clinical applications, there is manual intervention which is time consuming and needs most experienced physicians to reduce errors. By introducing automation in clinical applications which supports medical imaging workflow will make less time consumption for analyzing and feature extraction/segmentation. In last few decades, various methods have been introduced for classification of especially medical images, but typically, they perform well only on a specific subset of images, do not generalize well to other image sets, and have poor computational performance. MRI captures brain images in three modalities such as T1-weighted, T2-weighted, and proton density-weighted. A new wavelet-based method for the fusion of above three different channels in MRI image segmentation is used to produce a high contrast resulting image to increase the segmentation accuracy. This new segmentation step based on Machine learning concepts is introduced to train the machine for further MRI image segmentation towards a fully automated method. The ant colony optimization algorithm has a potential to incorporate human intelligence and prior knowledge about intensity and other tissue information, shape, size, symmetry, and normal anatomic variability and hence improves the segmentation results. Brain Web MRI image dataset with added noise is being used to compare with some known reported works like Fuzzy C means and Watershed algorithms and the obtained results are less immune to noise compared to ant colony optimization results. The segmented results are evaluated taking various performance parameters like Global Consistency Error (GCE), Rand Index (RI), and Variation of Information (VoI). It has been observed that there is a better performance in ant colony optimized segmentation images in terms of accuracy.

30.1   INTRODUCTION

Over the last few decades, the rapid development of noninvasive brain imaging technologies has opened new horizons in analyzing and studying the brain anatomy and function. Enormous progress in accessing brain injury and exploring brain anatomy has been made using magnetic resonance imaging (MRI). The advances in brain MRI have also provided large amount of data with an increasingly high level of quality. The analysis of these large and complex MRI datasets[10,11] has become a tedious and complex task for clinicians, who have to manually extract important information. This manual analysis is often time-consuming and prone to errors due to various inter- or intraoperator variability studies. These difficulties in brain MRI data analysis required inventions in computerized methods to improve disease diagnosis and testing. Nowadays, computerized methods for MR image segmentation have been extensively used to assist doctors in qualitative diagnosis.

From the rigorous review of related work and published literature, it is observed that many researchers have designed algorithms for the detection and segmentation of brain tumor from MRI images[1] by applying various techniques.[3] Comparative study of different segmentation techniques is summarized with advantages and disadvantages.

In general, the conventional medical image segmentation methods are classified into four categories:

- Threshold-based techniques

- Region-based techniques
- Pixel/Voxel classification methods
- Model-based methods

Threshold-based methods are the oldest and simplest methods which require a well-separated histogram. In medical images, it is rare to find a well-separated histogram; in general, all medical images will have overlapped regions with various features.[1] Even though it is the easiest method, one cannot use these threshold-based methods for medical image segmentation.

Region-based methods are also simplest segmentation methods which segments based on connectivity, that is, 4- and 8-point connectivity in the region of interest. Here, the regions with similar properties can be separated. Region growing and watershed are most widely used region-based segmentation methods but these are not using for medical segmentation due to partial volume effects, noise, and variation of intensity which lead to over segmentation.

Pixel/Voxel-based classification methods are mostly used for medical image segmentation. Most widely used classification methods are clustering methods, that is, fuzzy C means (FCM)[2] method which divides the image into different number of clusters with defined similarity and features. Here, different supervised and unsupervised methods are used for image segmentation.[7] In these methods, the computation time and controlling heuristic parameters will decide the efficiency and segmentation accuracy. Support vector machines gave better results for segmentation but it takes more time to execute and more time to train the machine. So, we should control all the heuristic parameters to improve the execution/computation time and segmentation accuracy.

Model-based techniques include various methods which model the parameters and pixel values into Gaussian and other parametric models which give better results but with increased complexity. These models may converge to wrong boundaries in case of nonhomogeneities. Level-set methods are also widely used segmentation technique[8,9] that needs a very good initialization and this initialization was done by FCM[4] method. If there is any mistake in initialization, it is prone to errors in segmentation. But all these methods are computationally expensive.

Based on the existing literature, several general conclusions[6] can be drawn with regard to elements of a system that can be used to improve performance in brain tumor segmentation.[2] So to address the segmentation toward a fully automated method, we are introducing the ant colony optimization (ACO) algorithms to incorporate human intelligence and prior knowledge about intensity and other tissue information, shape, size, symmetry, and normal anatomic variability to improve segmentation results.

For solving optimization problem, the ACO algorithm is used here. The unique characteristic of ants which is used in more applications is their behavior in the process of searching for their food. In particular, they search by finding the shortest path between the food source and their nest. Actually, there is no direct communication among the ants in the colony, but ants leave a chemical substance called a pheromone in the path as their indirect communication, and although the substance evaporates rapidly, for short periods of time, it remains and can be recognized on the ground as a trace of the ant and the path that they have taken. Basically, each ant chooses the path of greatest pheromone trace; in other words, an ant tracks the path that the most other ants have passed through and assumes that this most traveled path has the best source of food. This simple scheme is an effective mechanism for finding the optimal solution or best path selection. For cases in which convergence in value is the item to be considered, the algorithm is run to an optimal value.

30.2   PROPOSED SEGMENTATION METHOD

In this segmentation method, the segmentation of image into cerebrospinal fluid (CSF), gray matter (GM), and white matter (WM) was done in two stages. In the first stage, a wavelet (Haar transform)-based image fusion method was implemented to increase the image contrast and quality which favors segmentation of the image and in the second stage, ACO algorithm is implemented for segmenting the image and the segmentation was evaluated using different evaluation metrics like global consistency error (GCE), Rand Index (RI), peak signal-to-noise ratio (PSNR), and variation of information (VoI).

30.2.1   WAVELET-BASED IMAGE FUSION

Initially, different image modalities are divided into four different channels such as low horizontal and low vertical (LL) and the detailed images consist of high horizontal and low vertical (HL), low horizontal and high vertical (LH), and high horizontal and high vertical (HH) frequencies. This is explained below in stepwise manner.

Computation of Haar wavelet transform:

Step 1: Consider the rows of the image matrix and find the average and differences of each pair of pixels (we will get $n/2$ averages and $n/2$ differences).

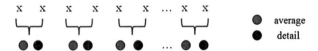

Step 2: Fill the first half of the array with averages and next half with differences of the above step as shown in the below figure.

Step 3: Now consider columns of rearranged matrix and repeat the Steps 1 and 2 for the columns also.

By the end of Step 3, we will get the first level of discrete wavelet transform as shown in Figure 30.1.

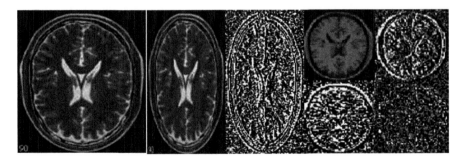

FIGURE 30.1   (a) Original, (b) after Step 2, and (c) after Step 3.

Step 4: Repeat the above steps for next level of transform.

Now perform the image fusion by considering T1 and T2 components in LL channel and T2 and proton density components[7] in remaining three channels, that is, LH, HL, and HH. The results of this fusion are shown in Figure 30.2.

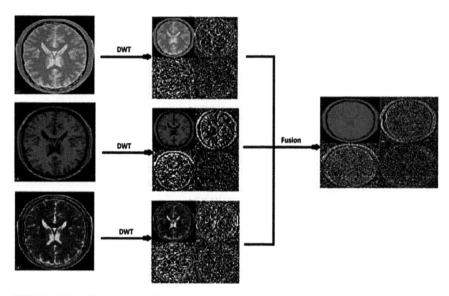

FIGURE 30.2   The results of fusion.

30.2.2   FUZZY C-MEANS ALGORITHM

FCM is a method of clustering which allows one piece of data to belong to two or more clusters. The FCM algorithm[5] is an unsupervised fuzzy clustering algorithm and is frequently used in pattern recognition. Conventional clustering algorithm finds "hard partition" of a given dataset based on certain criteria that evaluate the goodness of partition. By "hard partition," we mean that each datum belongs to exactly one cluster of the partition, while the soft-clustering algorithm finds "soft partition" of a given dataset. In "soft partition," datum can partially belong to multiple clusters. "A soft partition is not necessarily a fuzzy partition, since the input space can be larger than the dataset. However, most soft-clustering algorithms do generate a soft partition that also forms fuzzy partition. A type of soft clustering of special interest is one that ensures membership degree of point x in all clusters adding up to one, that is, "a soft partition that satisfies this additional condition is called a constrained soft partition. The FCM algorithm, which is best known fuzzy clustering algorithm, produces constrained soft partition. To produce constrained soft partition, the objective function J1 of hard C means has been extended in two ways."

The fuzzy membership degree in cluster has been incorporated in the formula.

$$\sum_j \mu_{cj}(x_i) = 1 \forall x_i \in X$$

An additional parameter m has been introduced as a weight exponent in fuzzy membership. The extended objective function, denoted by J_m, is

$$J_m(P,V) = \sum_{i=1}^{k} \sum_{x_k \varepsilon X} (\mu_{ci}(x_k))^m |x_k - v_i|^2$$

where P is fuzzy partition of dataset X formed by $C1, C2, \ldots, C_k$, and k is number of clusters. The parameter m is weight that determines the degree to which partial members of cluster affect the clustering result. Like hard C-means, FCM also tries to find good partition by searching for prototype v_i that minimizes the objective function J_m. Unlike hard C-means however, the FCM algorithm also needs to search for membership function μ_ci that minimizes J_m. A constrained fuzzy partition $\{C1, C2, \ldots, C_k\}$ can be local minimum of the objective function J_m only if the following conditions are satisfied.

$$\mu_{ci} = \frac{1}{\sum_{j=1}^{k} \left(\|x - v_i\|^2 / \|x - v_j\|^2 \right)^{1/(m-1)}}$$

Few important points regarding the FCM algorithm: It guarantees converge for $m > 1$. It finds local minimum of the objective function J_m. The result of applying FCM to a given dataset depends not only upon the choice of parameter m and c but also on the choice of initial prototype.

30.3 EXPERIMENTAL PART

30.3.1 *ANT COLONY OPTIMIZATION ALGORITHM*

Medical image segmentation using the ant colony optimization algorithm can be considered a process in which ants are looking for similar pixels (defined as food sources with specified features) by using vector features that are not identical. These food sources are considered as threshold limits for image segmentation, and the optimal value of this threshold limit is being acquired after implementation of the algorithm. The ACO algorithm runs automatically without the need for any manual operator interaction.

To begin the whole image is divided into $N\,N$ windows. It was experimentally determined by applying the ACO algorithm to three MR images of 1.5- and 3-TMR systems that setting N equal to 3 achieves excellent results and computationally is more efficient than when using larger window sizes such as 5×5 or 7×7.

All ants are propagated uniformly and randomly on the whole MR images pace (search space) to perform the search activity. For each of the targeted windows, a histogram curve is plotted based on the amount of pheromone trace, which for the case of application to medical image segmentation, is analogous to groups of image pixels containing all, some, or none of the object within the 3×3 search window. Thus, there are three possible scenarios for each window in the entire image:

1. The entire window falls in background (which is completely black).
2. The entire window falls in target.
3. The window falls at the boundary of target and in background.

In the first case in which the search window falls entirely within the image background, after plotting the histogram curve, no change is observed from any ant so that the search process is performed in this area just in the first iteration, and all the energy of ants is applied for target segmentation in the following iterations. So, as stated above, the entire ant's focus would be in the second mode. It should be noted that the novelty of our method is in the way that a specified ant is determined for each window and how the histogram curve is stored in its memory.

As previously mentioned, ants are placed randomly in the first pixel, and the intensity of that pixel and that pixel's surrounding or eight nearest neighboring pixels are stored in the ant's memory. After storing the intensity information of each window, the searchant shares its information with the ant that has the histogram curve in its memory. The most accurate threshold in intensity variation of neighboring pixels is determined from this master histogram.

This computation is done according to the following equations:

$$Ant_{i,t} = n_W\left(i,t\right)$$
$$n_W\left(i,t\right) = D \in W, \quad \left\{\left|D_i\right| \le W\right\}$$

where i identifies the ant that holds in memory the information about a desired pixel and its surrounding pixels in the selected window, t is the number of iterations in the process of implementing the algorithm, W represents the selected window for the search process in the image, and D is the number of pixels in the window W. "The constraint that $\{|D_i| \leq W\}$ assures that the ant in the window in the first iteration does not arrive into the next window until the process of segmentation in every window has been completed."

If we assume that C_i is the center of the ith window, "Ant$i$ places the ant in the C_i window, (a_i, b_j) is the pixel location of Ant$_i$, and (c_i, d_j) represents the position of ith window; under these conditions," each of the following equations must be met in each window:

$$\left\{|a_i - c_i| \leq \frac{N-1}{2}, \quad |b_i - d_i| \leq \frac{N-1}{2}\right\}$$

where N is the window size, a_i is the line up on which the ant is situated, c_i is the central line of the window, b_i is the column in which the ith ant is situated, and d_i is the central column of the window.

When the entire image is covered by searching ants and information for the whole image (intensity of the entire desired image pixels) is saved in the histogram-storing ant's memory, the "food sources"—which are analogous to the different types of brain tissue WM, GM, and CSF—are then defined by the optimum results of the ACO algorithm. After defining the "food" in the memory of ants, "they get involved in the task of finding pixels with the similar features to the food. A constraint is that the motion of each ant from one pixel to another is ruled by the law of transition probability, thus affecting the movement of other ants and can be expressed by the following equation":

$$P_{i,j} = \frac{F_{i,j}\left(\tau_{i,j}(t)\right)}{\sum_{i,j} F_{i,j}\left(\tau_{i,j}(t)\right)}, \quad \text{If } (i, j) \in I$$

where i represents all of the pixels in the image to be searched in the optimization process, and if $\{(i,j) \notin I\}$, it is concluded that $\{P_{i,j} = 0\}$. In the above equation, $\tau_{i,j}$ represents the amount of pheromone or the image intensity for a given pixel located at i and j per iteration. $F_{i,j}(T_{i,j}(t))$ is defined by the following equation:

$$F_{i,j}\left(\tau_{i,j}(t)\right) = \tau_{i,j}^{a}(t)\zeta_i^{\beta}v$$

where α and β are "weighting coefficients with the constraint that $\{\alpha, \beta > 0\}$, and ϑ is related to pheromone trail update and is defined as follows":

$$v = 1 - \theta \times \rho$$

where ρ is the "reduction rate of pheromone quantity as the search goes forward for f_0 and $\{\theta > 1\}$ where θ is a constant parameter. In the below equation, ζ_i represents the features of the image (e.g., intensity) and has a determinant role in the convergence rate for pixel segmentation. This value is calculated for each pixel of each window and is expressed with the following equation":

$$\zeta_i \frac{1}{N} \sum_i n_i$$

where N is the number of pixels in each window, n_i is threshold limit for each window, and ζ_i evaluates the predetermined desired feature of each image to be tracked or searched for, which in this case represents pixel intensity.

This probability equation for ant movement can be further simplified as

$$p_{ij}^k(t) = \begin{cases} \dfrac{\left[\tau_{ij}(t)\right]^\alpha \left[\eta_{ij}\right]^\beta}{\sum_{k \in \text{allowed}_k} \left[\tau_{ik}(t)\right]^\alpha \left[\eta_{ik}\right]^\beta} & \text{if } j \in \text{allowed}_k \\ 0 & \text{otherwise} \end{cases}$$

Typical values of α, β, η, and ρ are (these are from experimental results)

$$\alpha = 10; \beta = 0.1$$
$$\eta = 0.05; \rho = 0.1$$

After the algorithm optimization has converged and the final criterion has been verified, segmentation of the "image is complete and all of the pixels in the image are placed into one of three categories consisting of either WM, GM, or CSF. The ACO algorithm requires that the final result be achieved with minimal error as determined by comparing the manual segmentation of brain tissues by a neuroradiologist specialist," with the automated extraction of tissues by the computer ACO algorithm.

In addition, the Euclidean distance criterion must be met as

applied to each two neighboring pixels such that their intensities are compared as follows: if two neighboring pixels meet the minimum distance constraint and have the same pixel intensity (or have similarity in low tolerance), then they are considered as belonging to the same class (same tissue WM, GM, and CSF); but if the two neighboring pixels have a very different intensity, then they are considered as being located at the boundary of and belonging to different classes.

The distance criterion is defined as follows:

$$d_{i,j} = \sqrt{\left||p_i|^2 - |p_j|^2\right|}$$

where $d_{i,j}$ is the distance criterion and $\{p_i, p_j\}$ represents the intensity of two neighboring pixels i and j.

The ACO algorithm requires that the final result be achieved with minimal error as determined by comparing the manual segmentation of brain tissues by a neuroradiologist specialist, with the automated extraction of tissues by the computer ACO algorithm.

30.3.2 PERFORMANCE EVALUATION OF SEGMENTATION

Various researchers on image segmentation provided different valuation parameters that can be used for evaluating image segmentation techniques. These evaluation parameters are RI, VoI, GCE, boundary displacement error (BDE), segmentation accuracy, precision recall measure, convergence rate, mean absolute error, PSNR, hamming distance, local consistency error, structural similarity index measure, and entropy.

30.3.2.1 GLOBAL CONSISTENCY ERROR

The GCE measures the extent to which one segmentation can be viewed as a refinement of the other. If one segment is a proper subset of the other, then the pixel lies in an area of refinement, and the error should be zero. If there is no subset relationship, then the two regions overlap in an inconsistent manner. The formula for GCE is

$$GCE = \min\{\}$$

where segmentation error measure takes two segmentations S1 and S2 as input and produces a real valued output in the range [0:1] where zero signifies no error. For a given pixel p_i, consider the segments in S1 and S2 that contain that pixel.

30.3.2.2 RAND INDEX

RI counts the fraction of pairs of pixels who's labelings are consistent between the computed segmentation and the ground truth averaging across multiple-ground truth segmentation. The RI or rand measure is a measure of the similarity between two data clusters.

The RI has a value between 0 and 1, with 0 indicating that the two data clusters do not agree on any pair of points and 1 indicating that the data clusters are exactly the same.[11]

30.3.2.3 VARIATION OF INFORMATION

The VoI metric defines the distance between two segmentations as average conditional entropy of one segmentation given the other and thus measures the amount of randomness in one segmentation (www.wikipedia.org).

30.3.2.4 PEAK SIGNAL-TO-NOISE RATIO

It gives quality of image in decibels (dB) and is given as

$$PSNR = 20 \log_{10} \frac{255^2}{Rows \times cols}$$

30.3.2.5 CONVERGENCE RATE OR EXECUTION TIME

Convergence rate is defined as the time period required for the system to reach the stabilized condition. The lesser the execution time, the better is the segmentation technique.

30.3.2.6 BOUNDARY DISPLACEMENT ERROR

The BDE measures the average displacement error of one boundary pixels and the closest boundary pixels in the other segmentation.

$$\mu_{LA}(u,v) = \left\{ \frac{u-v}{L-1} 0 < u - v \right\}$$

Among all these parameters for medical image segmentation, generally we consider GCE, RI, and VoI. Convergence rate is defined from the number of iterations that an algorithm takes to complete the segmentation process. When we are going to implement it on any field-programmable gate arrays (FPGA) kit, execution time can be calculated in terms of clock pulses.

30.4   RESULTS AND DISCUSSION

To assess the performance of the ACO algorithm procedure, we used brain web database to generate simulated MR images with different slice thickness, nonhomogeneities, and noise values. All of the tests were processed in Qt Creator using OpenCV libraries. The MR brain image[4] segmentation simulation results are mentioned below for different simulated brain images of size 181 × 217.

30.4.1   SIMULATIONRESULTSFORANTCOLONYOPTIMIZATION ALGORITHM

The resulting segmented images are shown below for ACO algorithm. Here, the algorithm was tested on different brain MR images generated from simulated web database (Figs. 30.3 and 30.4).[1]

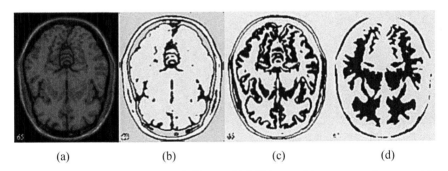

(a)   (b)   (c)   (d)

FIGURE 30.3   (a) Original image of Slice 75, (b) CSF image, (c) gray matter image, and (d) white matter image using ACO.

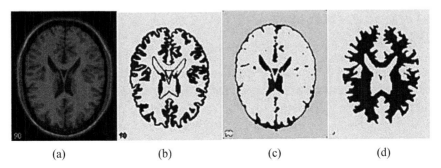

(a)   (b)   (c)   (d)

FIGURE 30.4   (a) Original image of Slice 90, (b) gray matter image, (c) CSF image, and (d) white matter image using ACO.

For all the above healthy brain images, the segmentation of brain image into CSF, GM, and WM was observed and segmentation accuracy was measured in terms of GCM, RI, and VoI (Figs. 30.5 and 30.6).[5]

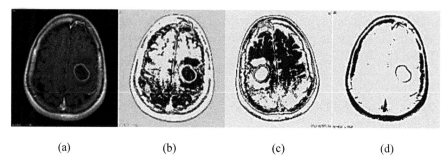

FIGURE 30.5   (a) Original brain image 1 with tumor, (b) gray matter image, (c) CSF image, and (d) white matter image using ACO.

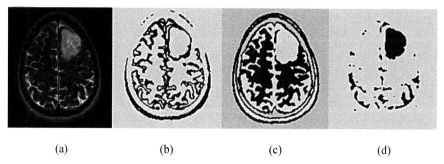

FIGURE 30.6   (a) Original brain image 2 with tumor, (b) gray matter image, (c) CSF image, and (d) white matter image using ACO.

30.4.2   PERFORMANCE EVALUATION OF SEGMENTATION RESULTS

Various researchers on image segmentation provided different valuation parameters that can be used for evaluating image segmentation techniques. These evaluation parameters are RI, VoI, GCE, BDE, segmentation accuracy, precision recall measure, convergence rate, mean absolute error, PSNR, hamming distance, local consistency error, structural similarity index measure, and entropy. The values of GCE, PSNR, and VOI are calculated for various images and some of them are shown in Table 30.1.

TABLE 30.1   Values of GCE, PSNR, and VOI for the images.

		ACO	FCM	Watershed
	GCE	0	0	0.0167
Image 1	Vol	0.8361	0.6659	0.8259
	PSNR	26.32	11.85	13.81
	RI	0.8463	0.9663	0.738
	GCE	0	0	0.0145
Image 2	Vol	0.8369	0.6562	0.7746
	PSNR	28.72	12.16	14.07
	RI	0.8234	0.9339	0.7134

30.5   CONCLUSION

In this work, a new algorithm was proposed based on real-time behavior of ants named ACO that is designed to achieve better results for the medical MR image segmentation with increased segmentation accuracy and to do so in a computationally efficient fashion. This algorithm was tested on different kinds of images and shown the segmentation results. From the experimental results, it is observed that the segmentation accuracy was improved by including wavelet-based data fusion method along with ACO algorithm. Present work can be extended on other MR images to improve segmentation results by reducing computation time and improving noise immunity. Further, this algorithm can be extended to different images like liver, heart, and chest MR images to extract the features of the images through segmentation. The segmented simulation results can be further analyzed with their volumetric analysis which is more useful for diagnosis.

KEYWORDS

- MR imaging
- MR image segmentation
- ant colony optimization
- brain image segmentation
- threshold-based techniques
- Haar transform
- soft-clustering algorithms

REFERENCES

1. Ahmadvand, A.; Kabiri, P. *Multispectral MRI Image Segmentation Using Markov Random Field Model*; Springer-Verlag: London, 2014.
2. Chitradevi, B.; Saranya, R. A Review on Brain Tumor Detection and Classification System Based on Image Processing Techniques. *Int. J. Sci. Res. Dev.* **2015,** *2* (12), 06–12, ISSN (online): 2321-0613.
3. Shinde, V.; Kine, P.; Gadge, S.; Khatal, S. Brain Tumor Identification Using MRI Images. *Int. J. Rec. Innov. Trends Comput. Commun.* **2014,** *2* (10), 3050–3054.
4. Taherdangkoo, M.; Bagheri, M. H.; Yazdi, M.; Andriole, K. P. *An Effective Method for Segmentation of MR Brain Images Using the Ant Colony Optimization Algorithm*; Society for Imaging Informatics in Medicine, April 2013.
5. Siyal, M.; Yu, L. An Intelligent Modified Fuzzy C-means Based Algorithm for Bias Field Estimation and Segmentation of Brain MRI. *Pattern Recognit. Lett.* **2005,** *26*, 2052–2062.
6. Gordillo, N.; Montseny, E.; Sobrevilla, P. State of the Art Survey on MRI Brain Tumor Segmentation. *J. Magn. Resonan. Imaging* **2013,** *31* (8), 1426–1438.
7. Kandwal, R.; Kumar, A. An Automated System for Brain Tumor Detection and Segmentation. In *Computational Intelligence and Computing Research (ICCIC), 2014 IEEE International Conference on*, Coimbatore, India, 18–20 Dec. 2014.
8. Fan, J.; Elmagarmid, Y.; Aref. In *Detection of Brain Tumor Using Threshold Value of MRI Images*. IEEE International Conference on Computational Intelligence & Computing Research (ICCIC), 2012, Coimbatore, India, Dec 18–20 2012.
9. Ali, S. M.; Abood, L. K.; Abdoon, R. S. Brain Tumor Extraction in MRI Images Using Clustering and Morphological Operations Techniques. *Int. J. Geogr. Inf. Syst. Appl. Remote Sens.* **2013,** *4* (1), 11–24.
10. Joseph, R. P.; Singh, C. S.; Manikandan, M. Brain Tumor MRI Image Segmentation and Detection in Image Processing. *Int. J. Res. Eng. Technol.* **2014,** *03* (01), 1–5. eISSN: 2319-1163, ISSN: 2321-7308.
11. Roy, S.; Bandyopadhyay, S. K. Detection and Quantification of Brain Tumor from MRI of Brain and Its Symmetric Analysis. *Int. J. Inf. Commun. Technol. Res.* **2012,** *2* (6), 477–483. ISSN 2223-4985.

CHAPTER 31

ADAPTIVE PILLAR K MEANS ALGORITHM TO DETECT COLON CANCER FROM BIOPSY SAMPLES

B. SAROJA* and A. SELWIN MICH PRIYADHARSON

Vel Tech Rangarajan Dr Sagunthala R&D Institute of Science and Technology, Avadi, Chennai, Tamil Nadu, India

Corresponding author. E-mail: boda.saroja@gmail.com

ABSTRACT

The subjective procedure in the analysis of tissue specimens in the grading of colon cancer depends mainly on the graphical assessment. In this chapter, an efficient method for detecting colon cancer from biopsy samples is presented. The first stage involves the formation of clusters in a set of redundant candidate region from the colon biopsy images using adaptive pillar K means algorithm. In the later stage, the tree structures are generated based on the information about outliers and each outlier acting as nodes. Finally, entropy-based outlier score computation is done on each node of the tree, then the score-based classification is performed to classify the colon biopsy images as normal or abnormal. The proposed method is implemented on Matlab working platform and the experimental results show that the proposed method has high achieved high classification accuracy compared with other methods.

31.1 INTRODUCTION

Colon cancer is one of the leading causes of cancer-related deaths in modern and industrialized world. About half a million people die every year worldwide due to colon cancer.[1] Colorectal cancer (CRC) (also known as colon cancer, rectal cancer, or bowel cancer) is the cancer of the large intestine (colon), the lower part of your digestive system. It arises from accumulated genetic and epigenetic alterations, which provide a basis for the analysis of stool to identify tumor-specific changes.[2] Most cases of colon cancer begin as small, noncancerous (benign) clumps of cells called adenomatous polyps. Over time, some of these polyps become colon cancers. The primary reason of colon cancer is chain smoking, but there are some other reasons of colon cancer, such as family history of colon cancer, increasing age, and unbalanced diet, like diets with low consumption of fruits/vegetables and heavy consumption of meat.[3] It is the development of cancer in the colon or rectum (parts of the large intestine). It is due to the abnormal growth of cells that have the ability to invade or spread to other parts of the body.[4]

Signs and symptoms may include blood in the stool, a change in bowel movements, weight loss, and feeling tired all the time. CRC develops and progresses as a consequence of abnormal cellular, molecular changes,[5] many of which result in mutant DNA. Modern molecular techniques allow examination of individual patient genetic data that ascribe risk, predict outcome, and/or modify an approach to therapy.[6] Screening for CRC is a highly effective intervention that substantially reduces cancer-specific mortality by detecting early-stage CRC and premalignant lesions.[7-9] A DNA stool test looks for certain gene changes that are sometimes found in colon cancer cells. Like other colon cancer-screening tests, it can find some colon cancers early, before symptoms develop, when they're likely to be easier to treat. Some screening tests can also sometimes find growths called polyps so they can be removed before they turn into cancer. That means screening can sometimes prevent colon cancer altogether.

Stool DNA (sDNA) testing has emerged as a feasible approach to CRC screening.[10] Its patient-friendly features have the potential to enhance screening compliance. It is noninvasive and requires no unpleasant and time-consuming cathartic preparation, no diet or medication restrictions, and no disruption of daily activities or work time and can be accessed by mail. Furthermore, sDNA testing has achieved highly accurate detection rates of curable-stage CRC and large advanced premalignant colorectal lesions.[11] Because DNA mutations may differ between colon cancers, sDNA tests typically target multiple markers to achieve high detection rates. Also, because DNA markers may be present in only trace quantities in stool, very sensitive laboratory methods are required. The new sDNA tests demonstrate high detection rates of early-stage colon cancer.

As a noninvasive CRC screening test, a multimarker first-generation sDNA (sDNA V 1.0) test is superior to guaiac-based fecal occult blood tests.[12] Unlike other noninvasive tests, the new sDNA tests also can detect precancerous polyps. CRC and advanced precancers can be detected noninvasively by analyses of exfoliated DNA markers and hemoglobin in stool.[13] An improved sDNA assay (version 2), utilizing only two markers, hyper methylated vimentin gene (hV), and a two-site DNA integrity assay (DY) demonstrated in a training set (phase 1a) an even higher sensitivity (88%) for CRC with a specificity of 82%. Thus, the multiple stool sampling practiced with fecal occult blood tests may not be necessary with sDNA tests.[14] The classical method of cancer detection is the microscopic inspection of colon biopsy samples; however, it is time consuming and laborious for the histopathologists and has interobserver/intra-observer variations in grading. Therefore, automatic colon cancer detection techniques are in high demand. Researchers have been working since decades to propose reliable automatic methods of colon cancer detection.[15]

In this chapter, a new approach for the detection of tumor in colon biopsy sample in the presence of outliers is developed. Here, we present a new efficient method for detecting the colon cancer in the presence of outliers. Initially, the colon biopsy samples are preprocessed using adaptive pillar K means clustering algorithm to produce set of redundant candidate regions in which clusters are formed. Then, the outliers within the clustered regions are generated as a tree structure in which the outliers are nodes, and the relationship between nodes is produced on the basis of information about outliers. The decision tree is used for tree structure generation. Then, entropy-based outlier score computation will be done on each node of the tree, which is obtained by an information gain (IG) method. Finally, score-based classification will be performed to classify the colon biopsy images as normal or abnormal.

The rest of the chapter is structured as follows. Section 31.2 illustrates the proposed technique and the description in detail. Section 31.3 provides experimental setup, results, and the associated discussion, performance measures which have been used for evaluation purposes. Section 31.4 concludes this research work.

31.2 EXPERIMENTAL PART

The colon is the large intestine and in the biopsy sample images, it is divided into three major parts: they are lumen and epithelial cells in white color, nuclei in purple color, and connecting tissues in pink color. Generally, the presence of colon cancer cells can be evident by investigating the lumen since that occurs in the inner wall of the colon. Sometimes due to the flaw in experimental procedures, there may be the chance of outliers which could degrade the performance of the cancer detection. In our proposed method, we present a new efficient method for detection of colon cancer in the presence of outliers. Initially, the colon biopsy samples are preprocessed using adaptive pillar K means clustering algorithm to produce set of redundant candidate regions in which clusters are formed. Then, the outliers within the clustered regions are generated as a tree structure in which the outliers are nodes, and the relationship between nodes is produced on the basis of information about outliers. Then, entropy-based outlier score computation will be done on each node of the tree. Finally, score-based classification will be performed to classify the normal or malignant cells. The overall system design of the proposed method is illustrated in Figure 31.1.

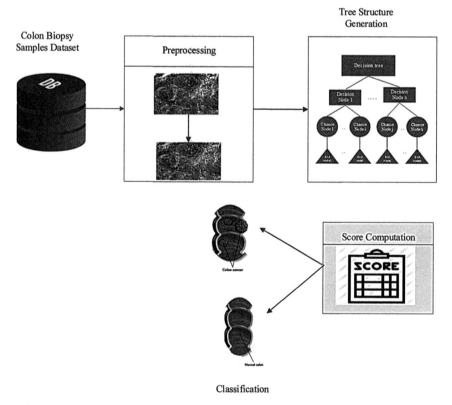

FIGURE 31.1   System design of proposed work.

31.2.1   IMAGE DATABASE

In this present chapter, the colon biopsy images are obtained from http://www.informed.unal.edu.co:8084/BiMed/. In these databases, we have taken totally 100 colon biopsy images represented by $di = \{d1, d2, \ldots dn\}$, where $n = 100$, out of that 85 images are used for training purposes and 15 images are used for testing purposes. In those 100 images, 70 are abnormal (cancer) images and 30 are normal images. The training images trained by forming clustering region then generate the decision tress and finally compute the entropy-based score and score-based classification to produce the result as normal or abnormal. The testing images are tested by the conditions apply to training image and produce the result as normal or abnormal. Figure 31.2 shows some of the normal and abnormal images from the data set.

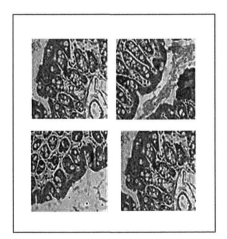 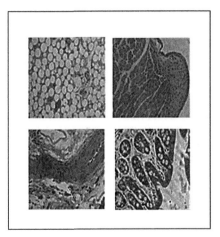

FIGURE 31.2   (a) Abnormal colon biopsy images from di and (b) normal colon biopsy images from di.

31.2.1.1   PREPROCESSING

Preprocessing is used to improve quality of the image and suppress the unwanted distortions or noise. The preprocessing step is usually performed to make the images fit for next phases, especially in the tree structure generation phase. In this research work, adaptive pillar K means algorithm is used to preprocessing such as to divide the colon biopsy images $d1$ to constituent clusters as Ck. $d1$ is usually characterized by pink-colored connecting tissues, purple-colored nuclei, and white-colored epithelial cells and lumen. Therefore, K means algorithm with $K = 3$ is applied to color intensities of pixels to divide an image pixel into three clusters.

31.2.1.1.1   Adaptive Pillar K Means Algorithm with the Efficient Distance Metric

The system utilizes the authentic size of $d1$ to perform high-quality image preprocessing which causes high-resolution image data points to be clustered. Therefore, the K means algorithm is utilized for clustering $d1$ data by considering that it has the ability to cluster immensely colossal data and additionally outliers payments are utilized expeditiously and efficiently. Because of starting points engendered arbitrarily, one of the local minima leads to erroneous clustering results so K means algorithm is arduous to reach global optimum. To evade this phenomenon, the proposed system utilizes adaptive pillar algorithm, which is very robust and superior for initial cluster optimization for K means by deploying all centroids far discretely among them in the data distribution. This algorithm is inspired by the cerebration process of determining a set of pillars' locations to make a stable house or building.

Locate two, three, and four pillars, to withstand the pressure distributions of several different roof structures collected of discrete points. It is inspiring that by distributing the pillars as far as possible from each other within a roof, as the number of centroids among the gravity, weight of data distribution in the vector space, the pillars can endure the roof's pressure and alleviate a house or building. Therefore, this algorithm designates positions of initial centroids in the farthest accumulated distance between them in the data distribution.

The adaptive pillar algorithm is described as follows. Let $\Delta C = \{dc \mid c = 1,...,n\}$ be colon biopsy sample, c be number of clusters, $A = \{ai \mid i = 1...c\}$ be initial centroids, $xy \subseteq \Delta C$ be identified of ΔC which is already selected in the sequence of processes, $AccD = \{g_i \mid i = 1,...,n\}$ be accumulated distance metric, $EucD = \{g_i \mid i = 1,...,n\}$ be Euclidean distance metric for each iteration, and M be the grand mean of ΔC. The execution steps of the proposed algorithm are described as follows:

ALGORITHM 31.1 Adaptive pillar K means.

Input: Colon biopsy sample dc, number of pillars C
Output: Optimized centroids A

Begin
Step 1. Set Z = Ø, SQ = Ø, and $AccD$ = []
Step 2. Calculate Euclidean distance
$EucD$ ← dis (dc,M)
Step 3. Set number of neighbors $N_{min} = \alpha\, N / k$
Step 4. Assign d_{max} ← arg$_{max}$ ($EucD$)
Step 5. Set neighborhood boundary
$Nb_{dis} = \beta \cdot d_{max}$
Step 6. Set $i = 1$ as counter to determine the ith initial centroid
Step 7. $AccD = AccD + (EucD)$
Step 8. Select ж ← gargmax(DM) as the candidate for ith initial centroids
Step 9. $xy = xy \cup$ ж
Step 10. Set EucD as the Euclidean distance metric between d_c and ж
Step 11. Set no ← number of data points fulfilling Nbdis EucD
¹Step 12. Assign AccD(ж) = 0
Step 13. If no < N_{min},
Go to step 8
Step 14. Assign EucD(xy) = 0
Step 15. $A = A$ U ж
Step 16. $i = i + 1$
Step 17. If $i \le k$,
Go back to step 7
Step 18. Finish, in which A is the solution as optimized initial centroids.
End

Apply K means clustering algorithm after getting the optimized initial centroids and then the position of final centroids is attained. Final centroids are used as the initial centroids for getting authentic size of d1 and then apply the d1 data point clustering utilizing K means which can able to amend clustering results and make more expeditious computation for the image clustering as c1. From the clustered region, the tree structure is generated based on the outlier that is explained in the next phase.

31.2.1.2 TREE STRUCTURE GENERATION

The outliers in $c1$ generate the tree structure that is based on the decision tree. A decision tree is one of the most popular methods for discovering meaningful patterns and classification rules in a data set. The decision tree is useful because construction of decision tree does not require any domain knowledge. It can handle hi-dimensional data. Their representation of acquired knowledge in tree form is easy to assimilate by users. A decision tree technique is achieved in two phases: tree building and tree pruning. Tree building is done in top-down approach; the tree is partitioning all the data into subsets that contain instances with similar values. In this process, the data set is traversed repeatedly. Tree pruning is done in bottom-up approach used to improve the classification accuracy of the classifier by minimizing overfitting. Overfitting in the decision tree is the cause of misclassification error. In this section, assume that all of the outliers have finite discrete clusters, and there is a single target node called the classification. Each node of the clusters of the classification is called a class. A decision tree consists of three types of nodes:

- Decision nodes—commonly represented by squares
- Chance nodes—represented by circles
- End nodes—represented by triangles

In which, the outliers are decision nodes, and the relationship between nodes is produced on the basis of information about outliers; each chance node represents an entropy-based score computation, and each end node represents the score-based classification. The topmost node in a tree is the root node. The entropy-based score computation is associated with a splitting criterion which is chosen to split the data sets into subsets that have better class "separability," thus minimizing the misclassification error. Once the tree is built from the training data, it is then heuristically pruned to avoid overfitting of data, which tends to introduce classification error in the test data. In this decision tree, the most important question is which of the outliers is the most influential in determining the classification and hence should be chosen first. Entropy measures or equivalently IGs are used to select the most influential, which is intuitively deemed to be the outlier of the lowest entropy (or of the highest IG). This learning algorithm works by (1) computing the entropy measure for each outlier, (2) partitioning the set of examples according to the possible values of the outlier that has the lowest entropy, and (3) for each that is used to estimate probabilities, in a way exactly the same as with the Naive Bayes approach. A typical decision tree is shown in Figure 31.3.

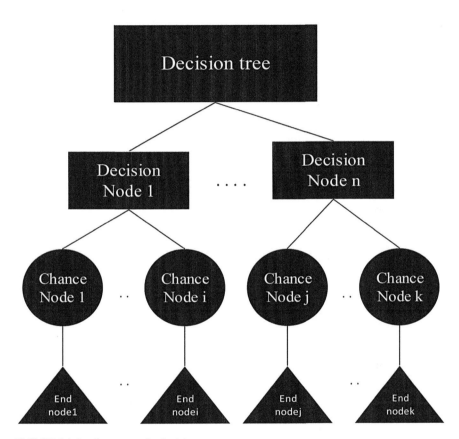

FIGURE 31.3   Structure of a decision tree.

Then, score will be computed for each and every node in the decision tree based on the score the colon biopsy images classified. The detailed explanation of score computation is present in the next phase.

31.2.1.3   ENTROPY-BASED SCORE COMPUTATION

Entropy-based score computation is used for classification purpose. From that entropy-based score, we have to easily classify that $c1$ is normal or abnormal. Entropy is commonly used in the information theory measure, which characterizes the purity of an arbitrary collection of $c1$. Entropy is also the measure of disorder or impurity. Entropy is the sum of the probability of each label times the log probability of that same label. In our case, we will be interested in the entropy of the output values of a set of training outliers. It is at the foundation of the IG-attributed ranking methods. The entropy measure is considered as a measure of system's unpredictability. The entropy of E is

$$A(E) = -\sum_{e \in \gamma} m(e) \log_2 (m(e))$$

(31.1)

where $m(e)$ is the marginal probability density function for the random variable E. If the observed values of E on the $c1$ are partitioned according to the values of a second information F, and the entropy of E with respect to the partitions induced by F is less than the entropy of E prior to partitioning, then there is a relationship between information E and F. Then, the entropy of E after observing F is

$$A\left(\frac{E}{F}\right) = -\sum_{f \in F} m(f) \sum_{e \in E} m\left(\frac{e}{f}\right) \log_2 \left(m\left(\frac{e}{f}\right)\right)$$

(31.2)

where $m(e/f)$ is the conditional probability of e given f. The $c1$ that goes down each branch of the tree has its own entropy value. We can calculate for each possible outlier its expected entropy. This is the degree to which the entropy would change if branched on this outlier. You add the entropies of the two children, weighted by the proportion of examples from the parent node that ended up at that child. The entropy typically changes when we use a node in a decision tree to partition the training instances into smaller subsets. IG is a measure of this change in entropy. Given the entropy as a criterion of impurity in an outlier S, we can define a measure reflecting additional information about E provided by F that represents the amount by which the entropy of E decreases. This measure is known as IG. It is given by

$$IG = A(E) - A\left(\frac{E}{F}\right) = A(F) - A\left(\frac{F}{E}\right)$$

(31.3)

IG is a symmetrical measure (refer to eq 31.3). The information gained about E after observing F is equal to the information gained about F after observing E. The info gain attributes scoring is subsequently used to create a neighboring molecule. IG is an entropy-based measure, which selects the node that has the best capability to differentiate the samples into separate classes. A node with a higher IG is considered more relevant. A neighboring molecule is generated by replacing entropy at a random position in an original molecule. During the process of tree construction, eq 31.3 is applied to find the score for the chosen node. This is done by scoring each node, using the purity entropy function and selecting the one that gives the optimum result.

The score is considered for every rule. The best set of rules is generated using the training data set. The score is well defined as follows:

$$S_N = \frac{NormalHit}{NormalCount} \qquad (31.4)$$

$$S_A = \frac{AbnormalHit}{AbnormalCount} \qquad (31.5)$$

$$Score = \begin{Bmatrix} S_N - S_A, \text{normal} \\ S_A - S_N, \text{abnormal} \end{Bmatrix} \qquad (31.6)$$

where *NormalHit* (*AbnormalHit*) is the number of normal (abnormal) samples that satisfy the relation of the rule among normal (abnormal) samples in a training dataset and normal count (tumor count) is the total number of normal (abnormal) samples in a training data set. The score ranges from 0 to 1.

S_N denotes the normal score and the S_A denotes tumor score. The difference between the S_N and S_A is a score for each rule. High-scoring rules can better differentiate the two classes. If S_N is greater than S_A for a given gene relation, then the rule's class label is normal. Otherwise, it is abnormal. Based on these scores, the $c1$ is classified as normal or abnormal and is explained in accompanying area.

31.2.1.4 SCORE-BASED CLASSIFICATION

Classification is the problem of identifying to which set of categories, a new observation belongs on the basis of a training set of data containing observations (or instances) whose category membership is known. When the best score is found, the corresponding model is output as the recognition result. A common way is to define a threshold for each model: if the score is better than the threshold we accept the result, otherwise reject it. But it is not easily applicable because speakers and utterances vary all the time and it requires quite some specialized knowledge and empirical information to get a good threshold. Furthermore, we can see that the threshold-selecting methods use only part of the score information (the best or the N-best). What if we use the entire score information? Here comes the idea. We don't care too much of the numerical value of a single score or several; on the other hand, we regard the scores on the speakers set (the set score) as a pattern of the speaker and try to apply a classifier on them. Obviously, it is a binary classification problem: to he or to he not. The score-based classification method is used to classify $d1$ as normal or abnormal. The scores are calculated based on the entropy in each node within the clustered region $c1$. First, the Shannon entropy is computed for $c1$; then, log energy entropy, threshold entropy, and sure entropy of $c1$ are computed. Finally, the norm entropy is computed and this is the score for the node. In score-based classification, eq 31.7 is shown that the entropy-based score value is reduced from the number of images; this is the major step for classification. Then, the obtained value is compared with the $c1$ threshold value.

$$X = N - r \qquad (31.7)$$

where X is the obtained value, N number of images, and r is the entropy-based score.

If the obtained value X is higher than the $c1$ threshold value, then $d1$ is said to be cancer (abnormal). If the obtained value X is lower than the $c1$ threshold value, then the colon biopsy image is said to be normal. Based on this technique, each and every colon biopsy image is classified.

31.3   RESULTS AND DISCUSSION

The proposed method is implemented in the working platform of MATLAB with the following system specification.

Processor   : Intel i5 @3 GHz
RAM   : 8 GB
Operating system   : Windows 8
MATLAB version   : R2013a

In this section, we have explained the experimental results in three sections: our proposed method result is shown in Sections 31.3.1–31.3.3, comparison of our proposed method with existing method is shown in Section 31.3.2, and the discussion is shown in Section 31.3.3.

31.3.1 DETECTION OF COLON CANCER IN PRESENCE OF OUTLIERS

31.3.1.1 PREPROCESSING

In this chapter, we are preprocessing the input image $d1$ using adaptive pillar K means algorithm. The adaptive pillar K means algorithm reduces the noise of $d1$. Then, clustering the redundant candidate region applies K means clustering algorithm after getting the optimized initial centroids and then attains the position of final centroids. Final centroids are used as the initial centroids for getting authentic size of the image and then apply the image data point clustering utilizing K means which can able to amend clustering results and make more expeditious computation for the image clustering. The clustering region calculates the threshold value that is used for classification purpose.

The sample input image $d1$ taken from the database for tumor detection is shown in Figure 31.4.

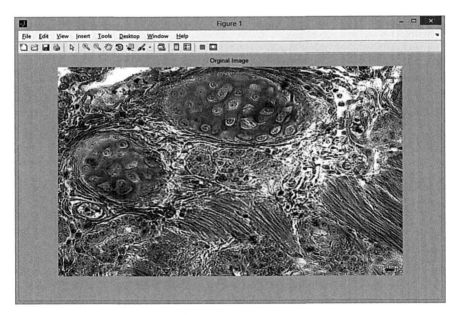

FIGURE 31.4 Input colon biopsy image.

The $d1$ shown in Figure 31.4 is used for adaptive pillar K means algorithm to reduce the noise; then, the redundant candidate regions formed into the cluster which is shown in Figure 31.5.

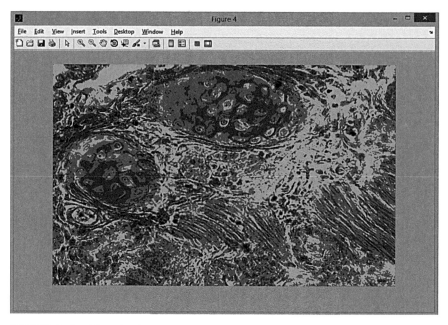

FIGURE 31.5   Clustering image.

From that $c1$, the outlier generates the tree structure based on the decision tree, in which the outliers are nodes and relationship between nodes based on the information about outlier. Then, entropy-based score computation is done on every node in the tree. An IG filter is used to compute the scores.

31.3.1.2   CANCER CLASSIFICATION

The outliers in the $c1$ generate the tree structure. On that, the entropy-based score is calculated for each and every node of the tree. The entropy is characterized by the purity of the given samples and also the measure of impurity. The IG method is used to calculate the score for each node. A node with a higher IG is considered more relevant. A neighboring molecule is generated by replacing entropy at a random position in an original molecule. During the process of tree construction, eq 31.3 is applied to find score for the chosen node. This is done by scoring each node, using the purity entropy function. The entropy score is then reduced from the number of images and it is compared with $c1$ threshold value. The obtained value is higher than the $c1$ threshold value; then $d1$ is said to be abnormal, otherwise normal. Finally, Figure 31.6 shows the output image for colon biopsy images.

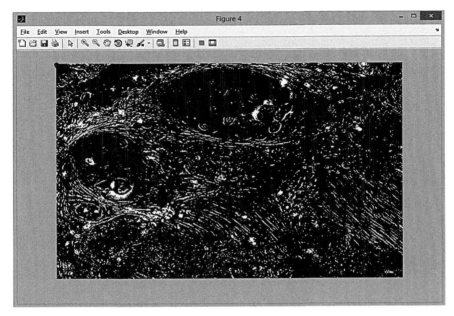

FIGURE 31.6   Output colon cancer image.

31.3.1.3   PARAMETER ANALYSIS

The classification capability of the various features proposed in this work has been quantitatively evaluated using various performance measures such as accuracy, sensitivity, specificity, Mathew's correlation coefficient, F score, and receiver operating characteristic curve. Normal and malignant images correspond to negative and positive samples, respectively. Therefore, true positive and true negative, respectively, are the number of correctly classified malignant and normal images. Similarly, false positive and false negative, respectively, represent the number of incorrectly classified normal and malignant images. The contingency table is given in Table 31.1.

TABLE 31.1   Contingency Table.

Actual class	Predicted class	
	Normal	**Abnormal**
Normal	TN	FP
Abnormal	FN	TP

31.3.1.1.1 Accuracy

The classification accuracy is a measure of usefulness of a technique. It depends upon the number of correctly classified samples and is calculated using the following equation:

$$\text{Accuracy} = \frac{Tp + TN}{N} \times 100 \tag{31.8}$$

where N is the total number of colon biopsy images.

31.3.1.1.2 Sensitivity

Sensitivity is a measure of the ability of a technique to correctly identify positive samples. It can be calculated using the following equation:

$$\text{Sensitivity} = \frac{TP}{TP + FN} \tag{31.9}$$

The value of sensitivity ranges between 0 and 1, where 0 and 1 mean worst and best recognition of positive samples, respectively.

31.3.1.1.3 Specificity

Specificity is a measure of the ability of a technique to correctly identify negative samples. It can be calculated using the following equation:

$$\text{Specificity} = \frac{TN}{TN + FP} \tag{31.10}$$

The value of specificity ranges between 0 and 1, where 0 and 1 mean worst and best recognition of negative samples, respectively.

31.3.2 COMPARISON BETWEEN THE PROPOSED METHODS WITH OTHER EXISTING METHODS

The performance of the proposed colon cancer diagnosis (CCD) system has been compared with previously proposed approaches of colon biopsy image classification. In this context, five techniques[16–20] have been selected from the contemporary literature for comparison. We have implemented these techniques in MATLAB and evaluated classification performance measures on the dataset described in Section 31.3.1.3. To obtain a fair comparison with proposed method, we have used optimal values of the parameters used in these techniques.

Table 31.2 represents the statistical measures of the proposed system for the given colon biopsy sample images.

TABLE 31.2 Statistical Measures of the Proposed System.

Sr. no.	Measures	Result		
		Proposed (%)	**Novel structural descriptor (%)**	**GECC (%)**
1.	Sensitivity	85.4	95.6	97
2.	Specificity	87.6	95.1	98
3.	Accuracy	99	95.40	98.67

The sensitivity value represents the percentage of recognition of actual value and specificity value represents the percentage of recognition of actual negatives. Accuracy is the degree of closeness of measurements of a quantity to its actual (true) value. The performance of the proposed CCD system is evaluated by comparing its classification results with a traditional classifier system which uses the novel structural descriptor- and gene expression-based ensemble classification of colon samples (GECC)-based tumor classification technique. Figure 31.7 represents the comparison graph of the statistical measure results of the proposed system with the novel structural descriptor- and GECC-based tumor classification system. The statistical graphs in Figure 31.7 show that the statistical measures give positive results for the proposed technique.

In Figure 31.7, the graph concludes that the sensitivity and specificity values of the proposed CCD system are lower than the existing method sensitivity and specificity value. The accuracy of the proposed method is higher than the existing method accuracy value, so the proposed method gives better performance.

To further prove that the proposed CCD system is the best for colon tumor detection, we made a comparison with some research papers which is shown in Table 31.3.

From the comparative analysis shown in Table 31.3, the proposed method has achieved better accuracy than the existing methods. From these experimental results, we can say that the proposed method is well suitable for the colon tumor identification scheme.

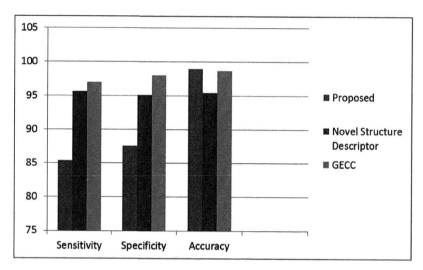

FIGURE 31.7   Comparison graph of the proposed system with the novel structure descriptor and GECC.

TABLE 31.3   Comparison Analysis with Previous Works.

Sr. no	Technique	Accuracy (%)
1.	Structural and statistical pattern recognition[1]	83.33
2.	Novel structural descriptors[3]	95.40
3.	Gene expression-based ensemble classification[15]	98.67
4.	Hybrid of novel geometric features[20]	92.62
5.	Proposed method	99

31.3.3   DISCUSSION

As far we have seen, there are several methods available for colon cancer detection, but due to the presence of outliers the quality of the detected output may degrade.. Like, in novel structure descriptor method of detecting the colon cancer, produces good result conversely the output gets affected due to the presence of outliers. Similarly, even though GECC is the most widely used methods of detecting colon cancer, its performance gets affected on producing the accurate detection result. But our proposed CCD method produces better performance due to the presence of outliers also. The comparison method has shown that the novel structure descriptor as well as the GECC method's sensitivity and specificity values are higher than the proposed method sensitivity and specificity values but the accuracy value is lower than the proposed method value. From these discussions, it can be stated that our proposed CCD method with adaptive pillar K means algorithm produces better results when compared with the existing detection methods in the presence of outliers.

31.4 CONCLUSION

Adaptive pillar K means algorithm to detect colon cancer from biopsy samples is presented in this chapter. In the proposed CCD method initially, the colon biopsy samples are preprocessed using adaptive pillar K means clustering algorithm to produce a set of redundant candidate regions in which clusters are formed. Then, the outliers within the clustered regions are generated as a tree structure based on the decision tree in which the outliers are nodes, and the relationship between nodes is produced on the basis of information about outliers. Then, entropy-based outlier score computation will be done on each node of the tree. The IG method is used to compute the score for the outliers. Finally, score-based classification is performed to classify the normal or malignant cells. Experimental results show that the proposed method has better results compared with existing methods. It further suggests that the proposed method is well suitable for the colon cancer identification scheme.

KEYWORDS

- **cancer biopsy samples**
- **adaptive pillar K means algorithm**
- **tree structure generation**
- **score computation**
- **CCD**
- **classification**

REFERENCES

1. Akbar, B.; Gopi, V. P.; Suresh Babu, V. In *Colon Cancer Detection Based on Structural and Statistical Pattern Recognition*. Proceedings of IEEE 2nd International Conference Electronics and Communication Systems (ICECS), 2015; pp 1735–1739.

2. Imperiale, T. F. Ransohoff, D. F.; Itzkowitz, S. H.; Levin, T. R.; Lavin, P.; Lidgard, G. P.; Ahlquist, D. A.; Berger, B. M. Multitarget Stool DNA Testing for Colorectal-cancer Screening. *N. Engl. J. Med.* **2014,** *370* (14), 1287–1297.

3. Rathore, S.; Hussain, M.; Iftikhar, M. A.; Jalil, A. Novel Structural Descriptors for Automated Colon Cancer Detection and Grading. *Comput. Methods Prog. Biomed.* **2015,** *121*, 92–108.

4. Keku, T. O.; Dulal, S.; Deveaux, A.; Jovov, B.; Han, X. The Gastrointestinal Microbiota and Colorectal Cancer. *Am. J. Physiol.-Gastrointest. Liver Physiol.* **2015,** *308* (5), G351–G363.

5. Leung, W. K.; To, K.-F.; Man, E. P. S.; Chan, M. W. Y.; Hui, A. J.; Ng, S. S. M.; Lau, J. Y. W.; Sung, J. J. Y. Detection of Hypermethylated DNA or Cyclooxygenase-2 Messenger RNA in Fecal Samples of Patients with Colorectal Cancer or Polyps. *Am. J. Gastroenterol.* **2007,** *102* (5), 1070–1076.

6. Carethers, J. M. DNA Testing and Molecular Screening for Colon Cancer. *Clin. Gastroenterol. Hepatol.* **2014,** *12* (3), 377–381.

7. Itzkowitz, S.; Brand, R.; Jandorf, L.; Durkee, K.; Millholland, J.; Rabeneck, L.; Schroy, P. C.; Sontag, S.; Johnson, D.; Markowitz, S.; Paszat, L.; Berger, B. M. A Simplified, Noninvasive Stool DNA Test for Colorectal Cancer Detection. *Am. J. Gastroenterol.* **2008,** *103* (11), 2862–2870.

8. Itzkowitz, S. H.; Jandorf, L.; Brand, R.; Rabeneck, L.; Schroy, P. C.; Sontag, S.; Johnson, D.; Skoletsky, J.; Durkee, K.; Markowitz, S.; Shuber, A. Improved Fecal DNA Test for Colorectal Cancer Screening. *Clin. Gastroenterol. Hepatol.* **2007,** *5* (1), 111–117.

9. Skally, M.; Hanly, P.; Sharp, L. Cost Effectiveness of Fecal DNA Screening for Colorectal Cancer: A Systematic Review and Quality Appraisal of the Literature. *Appl. Health Econ. Health Policy* **2013,** *11* (3), 181–192.

10. Kisiel, J. B.; Yab, T. C.; Nazer Hussain, F. T.; Taylor, W. R.; Garrity-Park, M. M.; Sandborn, W. J.; Loftus, E. V.; Wolff, B. G.; Smyrk, T. C.; Itzkowitz, S. H.; Rubin, D. T.; Zou, H.; Mahoney, D. W.; Ahlquist, D. A. Stool DNA Testing for the Detection of Colorectal Neoplasia in Patients with Inflammatory Bowel Disease. *Alimen. Pharmacol. Therap.* **2013,** *37* (5), 546–554.

11. Lidgard, G. P.; Domanico, M. J.; Bruinsma, J. J.; Light, J.; Gagrat, Z. D.; Oldham-Haltom, R. L.; Fourrier, K. D.; Allawi, H.; Yab, T. C.; Taylor, W. R.; Simonson, J. A.; Devens, M.; Heigh, R. I.; Ahlquist, D. A.; Berger, B. M. Clinical Performance of an Automated Stool DNA Assay for Detection of Colorectal Neoplasia. *Clin. Gastroenterol. Hepatol.* **2013,** *11* (10), 1313–1318.

12. Zou, H.; Taylor, W. R.; Harrington, J. J.; Hussain, F. T. N.; Cao, X.; Loprinzi, C. L.; Levine, T. R.; Rex, D. K.; Ahnen, D.; Knigge, K. L.; Lance, P.; Jiang, X.; Smitha, D. I.; Ahlquist, D. A. High Detection Rates of Colorectal Neoplasia by Stool DNA Testing with a Novel Digital Melt Curve Assay. *Gastroenterology* **2009,** *136* (2), 459–470.

13. Bosch, L. J. W.; Oort, F. A.; Neerincx, M.; Khalid-de Bakker, C. A. J.; Terhaar sive Droste, J. S.; Melotte, V.; Jonkers, D. M. A. E.; Masclee, A. A. M.; Mongera, S.; Grooteclaes, M.; Louwagie, J.; van Criekinge, W.; Coupé, V. M. H.; Mulder, C. J.; van Engeland, M.; Carvalho, B.; Meijer, G. A. DNA Methylation of Phosphatase and Actin Regulator 3 Detects Colorectal Cancer in Stool and Complements FIT. *Cancer Prevent. Res.* **2012,** *5* (3), 464–472.

14. Ahlquist, D. A.; Sargent, D. J.; Loprinzi, C. L.; Levin, T. R.; Rex, D. K.; Ahnen, D. J.; Knigge, K.; Peter Lance, M.; Burgart, L. J.; Hamilton, S. R.; Allison, J. E.; Lawson,

M. J.; Devens, M. E.; Harrington, J. J.; Hillman, S. L. Stool DNA and Occult Blood Testing for Screen Detection of Colorectal Neoplasia. *Ann. Intern. Med.* **2008,** *149* (7), 441–450.

15. Rathore, S.; Hussain, M.; Khan, A. GECC: Gene Expression Based Ensemble Classification of Colon Samples. *IEEE/ACM Trans. Comput. Biol. Bioinform. (TCBB)* **2014,** *11* (6), 1131–1145.

16. Wahaia, F.; Valusis, G.; Bernardo, L. M.; Almeida, A.; Moreira, J. A.; Lopes, P. C.; Macutkevic, J.; Kasalynas, I.; Seliuta, D.; Adomavicius, R.; Henrique, R.; Lopes, M. Detection of Colon Cancer by Terahertz Techniques. *J. Mol. Struct.* **2011,** *1006* (1), 77–82.

17. Tao, L.; Zhang, K.; Sun, Y.; Jin, B.; Zhang, Z.; Yang, K. Anti-epithelial Cell Adhesion Molecule Monoclonal Antibody Conjugated Fluorescent Nanoparticle Biosensor for Sensitive Detection of Colon Cancer Cells. *Biosens. Bioelectron.* **2012,** *35* (1), 186–192.

18. Kannen, V.; Garcia, S. B.; Silva, W. A.; Gasser, M.; Monch, R.; Joaquim Lopes Alho, E.; Heinsen, H.; Scholz, C.-J., Friedrich, M.; Heinze, K. G.; Waaga-Gasser, A. M.; Stopper, H. Oncostatic Effects of Fluoxetine in Experimental Colon Cancer Models. *Cell. Signal.* **2015,** *27* (9), 1781–1788.

19. Rathore, S.; Hussain, M.; Iftikhar, M. A.; Jalil, A. Novel Structural Descriptors for Automated Colon Cancer Detection and Grading. *Comput. Methods Prog. Biomed.* **2015,** *121* (2), 92–108.

20. Rathore, S.; Hussain, M.; Khan, A. Automated Colon Cancer Detection Using Hybrid of Novel Geometric Features and Some Traditional Features. *Comput. Biol. Med.* **2015,** *65* (1),1–17.

INDEX